THE CHEMISTRY OF FRAGRANCES

RSC Paperbacks

RSC Paperbacks are a series of inexpensive texts suitable for teachers and students and give a clear, readable introduction to selected topics in chemistry. They should also appeal to the general chemist. For further information on available titles contact:

Sales and Promotion Department, The Royal Society of Chemistry,
Thomas Graham House, The Science Park, Cambridge CB4 0WF, UK
Telephone: +44 (0) 1223 420066
Fax: +44 (0) 1223 423623

New Titles Available

Food – The Chemistry of its Components (Third Edition)
by T. P. Coultate
Archaeological Chemistry
by A. M. Pollard and C. Heron
The Chemistry of Paper
by J. C. Roberts
Introduction to Glass Science and Technology
by James E. Shelby
Food Flavours: Biology and Chemistry
by Carolyn Fisher and Thomas R. Scott
Adhesion Science
by J. Comyn
The Chemistry of Polymers (Second Edition)
by John W. Nicholson
A Working Method Approach for Introductory Physical Chemistry Calculations
by Brian Murphy, Clair Murphy and Brian J. Hathaway
The Chemistry of Explosives
by Jacqueline Akhavan
Basic Principles of Inorganic Chemistry – Making the Connections
by Brian Murphy, Clair Murphy and Brian J. Hathaway
The Chemistry of Fragrances
compiled by David Pybus and Charles Sell

Existing titles may be obtained from the address below. Future titles may be obtained immediately on publication by placing a standing order for RSC Paperbacks. All orders should be addressed to:

The Royal Society of Chemistry, Turpin Distribution Services Limited,
Blackhorse Road, Letchworth, Herts SG6 1HN, UK
Telephone: +44 (0) 1462 672555
Fax: +44 (0) 1462 280947

RSC Paperbacks

THE CHEMISTRY OF FRAGRANCES

Compiled by
DAVID PYBUS and CHARLES SELL

Quest International
Ashford, Kent, UK

ISBN 0-85404-528-7

A catalogue record for this book is available from the British Library

Published by The Royal Society of Chemistry,
Thomas Graham House, Science Park, Milton Road, Cambridge CB4 0WF, UK

For further information see our web site at www.rsc.org

Typeset by Paston Prepress Ltd, Beccles, Suffolk, NR34 9QG
Printed in Great Britain by Redwood Books Ltd, Trowbridge, Wiltshire

Preface

Modern perfumery is a blend of art, science and technology. Chemistry is the central science involved and the modern perfumery company houses specialists in all branches of chemistry, from physical chemists through analysts and synthetic organic chemists to biochemists. Indeed, the spectrum of skills ranges right across into biological fields, such as toxicology and sensory and behavioural sciences. All of these specialists work together with perfumers, and are essentially creative artists, accountants, marketeers and salespeople. No one person's skill is sufficient to meet the needs of the business and all must pull together as a team.

Our primary aim in this book is to show the use of chemistry in an exciting and rewarding business environment. However, we also felt that we should attempt to convey the interdisciplinary teamwork that is essential for success. To do this, we have invented a story that runs through the chapters and, hopefully, links them together in a way that shows how different specialists work together in a modern perfumery company.

The story begins when a fragrance house receives a brief from a customer. Fragrance houses do not sell directly to the public but to customers who manufacture consumer goods. These customers are those companies whose names are well known to the general public. Some may sell only fine fragrances, others may specialize in household cleaners. Yet others manufacture a broader spectrum, perhaps even the entire range from fine fragrances to industrial detergents. The customer in our story falls more into the last category and the brief asks for a perfume that can be used in a fine fragrance, but could also be 'trickled down' into a toilet soap, a shampoo and an antiperspirant. Such a request puts considerable constraints onto the creative perfumer, as described herein, and the combined efforts of perfumer and chemists are necessary to provide a suitable fragrance.

We hope that the story element adds to the enjoyment of the reader and gives a deeper understanding of the fascinating world of perfumery than would a simple factual account of the subject matter.

David Pybus
Charles Sell

Contents

Glossary

ABSOLUTE The alcoholic extraction of the concrete.

ACCORD A blend of perfume ingredients balanced in odour intensity and having a pleasing effect. Generally used as a perfume building block.

ALCOHOLIC A perfumed product designed for application to the skin in which the carrier used is aqueous ethanol, *e.g. Chanel N°5.* Concentration of perfume can vary from *ca.* 2% for an eau de cologne to *ca.* 30% for an extract.

ALDEHYDIC An odour descriptor used to define the effect of using relatively large amounts of aliphatic aldehydes, *e.g.* C10 aldehyde.

ANOSMIA Inability to perceive an odour generally or specifically for certain molecules. Can be genetically inherited, be induced by drugs or be the result of damage to the nose or brain.

BALANCE A combination of perfume notes such that no particular note dominates the others.

BASE Can be a confusing term as it is used in two ways in the perfume industry: (1) to define a perfume sub-unit or building block (unlike an accord, a perfume base is balanced in terms of top-, middle- and end-notes) or (2) to define the unperfumed medium, *e.g.* soap noodles.

BOTTLE-NOTE The perception of the perfumed product on opening the bottle closure. Consists of perfume plus chemicals used in base (*e.g.* shampoo) manufacture.

BRIEF Document provided by the customer defining their perfume requirements.

CONCRETE The hydrocarbon extraction of the plant material.

END-NOTES The substantive part of a perfume comprising the less volatile components of the fragrance composition. Often crystalline, resinous or of high relative molecular mass (low volatility) liquid.

ESSENTIAL OIL The steam distilled oil obtained from plant material.

EXPRESSED A cold process in which the oil contained in the outer skin of a citrus fruit is released by rasping or compression of the citrus fruit, *e.g.* lemon, orange, bergamot. Sometimes known as 'cold pressed' oil. Citrus oils would degrade if a normal steam distillation process was used.

FIXATION A fixative is used to prolong the effect of the more volatile ingredients in a perfume formula in an attempt to equalize the rate of evaporation of the component ingredients. Molecules with low vapour pressure are used; these are often resinous, crystalline or high relative molecular mass liquids.

FORMULA The list of ingredients with their proportions required to produce the desired odour effect.

HEADSPACE Headspace is the air above or surrounding a fragrant substance which contains volatile compounds. Any form of analytical procedure which samples and analyses this vapour is termed 'Headspace Analysis'.

HEDONIC Concerning the pleasurable sensation associated with fragrance.

LOG *P* The logarithm (base 10) of the octanol–water partition coefficient of a molecule.

MACERATION Extraction with hot fat. This process used to be applied to those flowers which gave a very small yield by distillation.

MIDDLE-NOTES The heart of a perfume; the main theme. Lasts for a few hours on the skin.

MUGUET Lily of the valley.

NATURAL PRODUCTS Plants that can provide odiferous materials, *e.g.* rose, sandalwood, grapefruit, *etc.*

NATURAL Perfume materials of natural origin. Derived directly from nature.

NATURE-IDENTICAL Perfume ingredients identical to those that occur in nature, but derived synthetically.

OSMOPHORE For a series of active molecules detected by the same mechanism, it is assumed that there is a common conformation in which key atoms or functional groups are placed at certain relative distances from one another. The spatial arrangement is known as a biophore and, specifically, an osmophore in odorant perception.

SCHIFF'S BASE Organic compound formed by reaction between an aldehyde or ketone with a primary amine, *e.g.* hydroxycitronellal forms a Schiff's base with methyl anthranilate to give Aurantiol®.

SMELLING BLOTTER Thin strip of highly absorbent paper used to assess the effect of a fragrance as it evaporates from top-note through to the end-notes. Also known as 'smelling strip' or 'mouillette'. One end of the blotter is dipped into the straight perfume oil or the oil diluted in alcohol.

SYNTHETIC INGREDIENTS Those materials obtained synthetically that are not identical to those found in nature, *i.e.* neither natural (*q.v.*) nor nature-identical (*q.v.*).

TOP-NOTES The most volatile components of a perfume. Generally lasts a matter of minutes on the skin.

TRANSPARENCY A general term for the trend towards lighter, fresher fragrances.

TRICKLE-DOWN The fragrance oil used in the alcoholic variant in a range is adapted for use in other variants in the range, such as soap, antiperspirant, *etc.* This usually involves cheapening and substitution of perfume ingredients for stability and performance.

TROPICAL Used as an adjective to describe the type of fruit (*e.g.* mango) or flower (*e.g.* ylang-ylang) found in the tropics. Sometimes also described as exotic.

WHITE FLOWERS Flowers with white petals, such as jasmine, tuberose, gardenia. Often contain indole, which causes the characteristic browning of the petals with time.

Chapter 1

A Home Full of Fragrance

DAVID PYBUS

Perfume comes in crystal bottles. Fact? Well, only partly. Personal perfume comes in crystal bottles. Tiny and elaborate, they are the modern expression of an ancient craft and precious resource that remain very relevant to life today. Perfume has become an intrinsic part of our lives; it may wake us up, sooth and comfort us. It is a part of our identity.

Think about an ordinary day and all the different smell sensations; the zesty, invigorating shower gel, the familiarity of a personal perfume, the fresh-washed smell of clean clothes, the citrus tang of the dish-wash liquid, the powdery, soft smell of an infant's skin ... the relaxing oil in your night-time bath.

The fragrance in each product we use is taken for granted, though (like a lot of other things that are taken for granted in our modern, sophisticated world) behind the scenes a whole industry strives constantly to improve fragrances; to make them more effective, longer lasting, and relevant to the values of the brand. The artisan perfumers work with precisely crafted fragrance materials supplied by chemists and designed to enhance modern products, both in terms of smell and function.

People are essentially visually oriented, and dependent on sight and sound to gather information from the surroundings. Smell, however, is an extraordinary sense. Closely linked to the limbic system (seat of emotions and the functions of memory), it has the power above all other senses to transport us, in an instant, to times past or pervade our psyche to change our mood. Only now is science starting to understand how this sense works, and scientists are discovering that it may be the most complex sense of all.

The consumer is ahead of the scientist, however. Now, more than ever before, the developed world is awash with products to enhance every aspect of modern living. The consumer is spoilt for choice, but a choice must be made! Fragrance is an important part in the positioning of these products and is a feature that the consumer turns to automatically to underscore the promise.

Fragrance is much more than personal perfume. It pervades every aspect of modern life, every different type of product. In a world where jobs are more demanding and less secure, crime rates are up, change is rife and traditional roles disappearing, products are appearing across the globe to sooth and reassure:

—room fragrances, from traditional candles to technology driven electrical 'plug-ins';
—bath products that promote aromatherapy have moved from luxury-brand goods to everyday goods sold by supermarkets;
—household cleaners and laundry products.

People may be visually oriented, but unconsciously we turn to other senses to simplify and de-stress our complicated existence.

Fragrance is mysterious, ethereal and elusive. Yet it is rooted solidly in the physical world and can therefore be examined scientifically. The chemistry behind fragrance is complex and fascinating. How do you build fragrance molecules to withstand heat and water and to emerge from the wash cycle firmly affixed to clothes, not washed away as the machine drains itself, and so convey messages of perfumed reassurance to the wearer? Perhaps this book will help to explain.

Chapter 2

The History of Aroma Chemistry and Perfume

DAVID PYBUS

In chemistry also, we are now conscious of the continuity of man's intellectual effort; no longer does the current generation view the work of its forerunners with a disdainful lack of appreciation; and far from claiming infallibility, each successive age recognizes the duty of developing its heritage from the past.

August Kekule von Stradonitz (1829–1896)

The discovery, exploitation and use of fragrant materials began with an elite few and had religious connotations. The very word 'perfume' is derived from the Latin *per fumum*, meaning 'by' or 'through' smoke, as it was with the use of burning incense that the prayers of the ancients were transported to the heavens for the contemplation of the Gods. Then came the priest-kings, and a wider audience, though still very select, of pharaohs, emperors, conquerors and monarchs with their attendant courtesans and alchemists, when use of perfume took on a hedonistic mantle as well as a spiritual one. By the twentieth century the combination of chemistry and the industrial revolution brought the revelation of perfume to the rest of humankind.

The great world religions of Islam, Christianity, Buddhism, Hinduism, Shintoism and Zoroastroism employ fragrance in pursuance of their faiths. Thus, religious and pleasurable pursuits have been the main drives in the phenomenal growth of perfume usage throughout the centuries. With the dawning of civilization, the use of fragrances developed within the four great centres of culture in China, India,

Egypt and Mesopotamia, and was extended in the sophisticated societies of Greece, Palestine, Rome, Persia and Arabia.

The seven ages of 'aromatic' man in Western culture began when Crusaders brought back three magical gifts from the East to the Dark Ages of Europe, which had 'not bathed for a thousand years'. Delicate aromatics, distilled alcohol and refined glass were the physical manifestations of thousands of years of alchemical research. The three together, a beautiful smell, a solvent to extend it and a bottle to conserve this 'gift from the Gods' were gladly accepted in the medieval West, and their use blossomed through the six ages of Chivalry, Alchemy, Discovery, Revolution, Empire, and Fashion.

EARLY USE OF FRAGRANCE

In prehistoric times, the hunter–gatherer tribes, in their explorations of nature, found many wonderful substances of extensive use in everyday living. Animal products in great variety, by-products of the hunt, were employed for clothing, shelter and tools, as well as for food. Similarly, the collection of herbs, spices and grasses unearthed familiar and fragrant compounds that were put to good use by the clans. An elite few appear to have been given special reverence to hold in trust the lore of the tribe. These sorcerers, or medicine men, knew the power, use and misuse of nature's pharmacopoeia, and over the centuries, by word of mouth, their store of wisdom increased. Craftsmen and artisans developed new and varied uses for materials as the human drive to extend and expand knowledge knew no bounds.

Eventually, a drift and concentration of tribes founded the great civilizations of the Nile in Egypt, Mesopotamia (between the Tigris and Euphrates) in modern-day Iraq, the Hwang-Ho valley in China and the Indus of Mohenjo Daro and Harappa, all of which came into their own between 4000 and 2000BC. Within these civilizations, over the centuries, knowledge of glass, alcohol and aroma chemicals was developed. Mesopotamians and Egyptians discovered that, when sand and ashes were heated together, a hard, brittle, transparent substance was produced. The addition of limestone hardened the glass and gave it more durability. Thus, from SiO_2, Na_2CO_3 and $CaCO_3$, was formed the vessel to hold a yet greater chemistry.

The first alcoholic wines were most likely an accidental discovery from fermented grapes. Most fruits are contaminated with microbes that form the surface bloom, which results in a natural alcoholic fermentation when given the right climatic environment. Earliest references to the production of distilled spirits appear to have originated in China around 1000BC, and it is believed that the production of beer developed in Egypt some 7000 years ago, involving the hydrolytic breakdown of starch in cereal extracts. Thus, with the earliest production of ethanol, we have another key compound in our fragrant mix.

Meanwhile a myriad of fragmented clues hint of ancient olfactory indulgence. Incense statuettes thousands of years old have been unearthed in the ruins of the Indus civilization, which was known to trade with both Egypt and Mesopotamia, while in China, around 500BC, Confucius proclaimed that both incense and perfume mitigate bad smells. Herbs, spices and flowers were used to ward off evil spirits, and flower-strewn graves over 5000 years old have been discovered in Iraq. In Mesopotamia, the fabled 'Garden of Eden' fragrant wood was used to build temples, and the fine smelling essences of cedarwood, myrtle and calamus reeds (sweet flag) were offered up for the pleasure of the Gods. It was here that the classic techniques of pressing, maceration and enfleurage, discussed in detail in Chapter 3, were developed. In supplication to the God Marduk, Nebuchadnezzar II, King of Chaldea, announced:

> *I anoint myself everyday with oil, burn perfumes and use cosmetics that make me worthier of worshipping thee.*

Early records detail King Sahure's trip to the fabled land of Punt, believed to be modern day Somalia or Ethiopia, around 2400BC, bringing back, amongst other riches, 8000 measures of myrrh. Temple pictographs dated around 1500BC detail Queen Hatshephut's journey to Punt, which had the objective of bringing back frankincense trees to replant in Egypt.

Pictographs show courtesans wearing bitcones on their heads, consisting of animal fat impregnated with aromatic materials. In the Egyptian heat this fat melted down the neck, covering the body in an oily, pleasant layer which, whilst workable as an early form of deodorant, could have caused much inconvenience to the wearer, and is a far cry from today's modern shampoo. Other hieroglyphs depict the great Ramses of Egypt offering incense in thanks to the Gods, while Nefertiti joined Semiramis of Babylon as one of the earliest women to demonstrate the liking of particular products, such as honey and orchid leaf, in her fragrant formulations. Perfume concoctions appear on the wall of the temple of Horus, at Edfu, amongst which is Kyphi, or Kephri, the 'twice-good' fragrance, burnt in the early morning and at eventide. A listing of the key ingredients demonstrates a sophistication in the formula at this time, and the wealth of products used. The formula of Kyphi contained spikenard, another prized material, a calcite vase of which was found by Howard Carter in 1922 at King Tutankhamen's tomb. Two of the most powerful and earliest reputed alchemists, Hermes Trismegitus and Zosimus, hailed from the land of the Nile, and did much to set down the records of their arts as future generations to develop.

The Christian Bible is chock-full of fragrance descriptions, from the early days of the tribes of Israel, when Joseph's brothers sold him as a slave:

> *A company of Ishmaelites came from Gilead with their camels bearing spicery and balm and myrrh, going to carry it down to Egypt.*

In Exodus, God gives Moses instructions for a holy perfume for himself, and a different one for his priests, whilst the Queen of Sheba's visit to Solomon was motivated by her wish to keep open the trade routes of the Arabian peninsula, her source of frankincense and myrrh, through Palestine to Egypt and Mesopotamia.

The story of Jesus of Nazareth is populated by fragrant materials, from frankincense and myrrh, his gifts at birth, through to the use of spikenard to wash his feet during life and finally the use of myrrh in the binding sheets of his body after crucifixion. Through trade and cultivation, Palestine became a great source of aromatic wealth, and a key trade route for myrrh and frankincense.

The Greeks further developed the use of fragrances, not only in praise of their Gods, but also for purely hedonistic purposes and for use in exercises and games, the first beginnings of early forms of aromatherapy. Their myths are full of references to aromas. Tear-shaped drops of the resin myrrh were the tears of a girl transmuted into a tree by the Gods. The hyacinth flower grew from the blood of dying Hyacinthus, struck by a discus during a feud between two other Gods. The iris grew at the end of a rainbow, whilst the narcissus flower grew at the spot near a mountain pool, where its erstwhile namesake drowned. Whilst a special fragrance formulation for the Goddess Aphrodite created such sensual desire that the term 'aphrodisiac' was used in its praise.

The sciences of medicine and herbalism developed with Hippocrates and Theophrastus, whilst Alexander the Great, tutored by Aristotle, conquered half the known world, acquiring a love of fragrance from the defeated Persian kings. But it was Aristotle who, in the third century BC, arguably advanced the cause of alchemy. It was he who observed the production of pure water from the evaporation of seawater. He

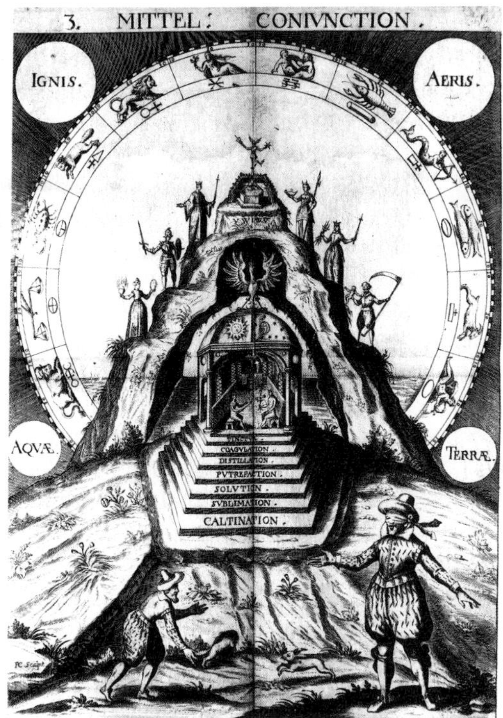

Figure 2.1 *The four elements, fire, air, water and earth*

translated the *Book of Hermes*, written by an Arab, Al-Makim. It expounded the theory, first suggested by Empedocles around 450BC, that all substances are made of the four elements, fire, earth, air and water (Figure 2.1). By varying the amounts of the different elements in each compound, all other elements could be made.

The theory developed further to discuss moods related to the elements and the seasons, as illustrated in Figure 2.2. The four key moods described were phlegmatic (solid, calm, unexcitable),

Hot	Summer	Dry	Autumn
Fire	**Choleric**	Earth	**Melancholic**
Wet	Spring	Cold	Winter
Air	**Sanguine**	Water	**Phlegmatic**

Figure 2.2 *Elements and moods*

choleric (irascible, hot-tempered), sanguine (optimistic, confident) and melancholic (sad, pensive). Combinations of the four moods at their boundaries give eight mood poles, and it is around these that some modern-day theories of aromatherapy have evolved.

The most used fragrances of the Greeks were rose, saffron, frankincense, myrrh, violets, spikenard, cinnamon and cedarwood, and to obtain these aromatics they traded far and wide throughout the Mediterranean and Middle East.

Meanwhile, in Rome, Pliny the Elder outlined a primitive method of condensation which collected oil from rosin on a bed of wool, and also made the first tentative experiments in chromatography. The Romans had developed techniques of enamelling, and made one of the most fundamental discoveries: that glass could be blown. The Roman contribution to perfumes consisted mainly in making an industry of the supply of raw materials and production of a large variety of fragrances in different forms. Military conquests secured new sources and supply routes to fit the steady demands of a far-flung empire, and the key products in demand were:

—Hedysmata: solid unguents, normally in the form of gums and resins;
—Stymata: liquid toilet waters infused with flower petals;
—Diapasmata: powdered perfumes using aromatics disposed in talc or gypsum.

Roman elite kept *Acerra*, small incense caskets, in their homes, and carried ampullae, perfume containers, and strigils, wooden blades for scraping oils off the skin at the hot baths. Petronius wrote:

Wines are out of fashion, Mistresses are in
Rose leaves are dated
Now Cinnamon's the thing.

The first professional perfumers (*unguentarii*) plied their trade in Capua, which became a trading centre of the industry. Perfume was used in abundance at the Games, both as a present for the crowds, and as a mask for the malodours of a bloodstained and offal-dappled arena. It is estimated that in the first century AD Romans were consuming nearly 3000 tonnes of frankincense and over 500 tonnes of the more expensive myrrh. Roman Emperors, of course, used perfume to excess, instanced by Nero and his wife Poppaea, who had a kind of 'perfumed plumbing' in their palaces, with false ceilings designed to drop flower petals onto dinner guests and scented doves which fragranced the air with their perfumed wings. When Poppaea died, it was said of Nero that he burned a whole year's supply of incense on her funeral pyre. A fragrant fortune which would have amounted to hundreds of tonnes. Towards the end of the Empire, Heliogabalus showed the true excesses of wealth and power. According to contemporary accounts he sported gilded lips, henna-dyed hands and feet and eyes decorated in concentric rings of blue and gold. This Emperor of Rome hailed originally from Syria, Land of Roses, which, as with Nero, were his favourite blooms.

When Rome succumbed to the barbarian hordes, the lights went out in all the incense burners throughout Europe, and the rose petals went out with the bath water.

THE AGE OF CHIVALRY

Whilst Crusaders became the implacable foes of Islam in the Holy Land throughout the eleventh and twelfth centuries, they admired many of the material possessions of Muslims, and brought back to their dank, dark and gloomy castles in Europe wall hangings, carpets, spices, eating forks, glass vessels and fragrances. Empress Zoe, in the Christian stronghold of Constantinople, had employed court perfumers, certain that incense and perfumes drove out demons. From there the practice spread, with Normans strewing flowers and rushes onto the floors of castles and churches to keep the air fragrant and acceptable.

It was common to employ a washerwoman, or lavenderess (from which the word 'laundress' is derived) to place sprigs and sachets of lavender around the rooms, and sweet-smelling packets of herbs amongst the bed linen. Not so pleasant odours were important too. Knights jousting for a lady's favour were not after a pretty handkerchief, but a 'pretty' smell (that of the lady's armpit odour), since there was a practice of holding a kerchief there to retain some of the smell, and remembrance of the wearer.

THE AGE OF ALCHEMY

In a perverse way, the Black Death of 1347–1351 and subsequent pandemics were huge catalysts to the growth in usage of aromatic products, which had already shown signs of flourishing from Eastern alchemical practice. Plague was believed to be caught by breathing foul air. Dead bodies lying in the street gave off the odour of decay, and to counteract this people carried nosegays and small floral bouquets (posies) from whence the ring-a-ring of roses children's nursery rhyme derived (red rings being a primary visual symptom of plague). To visit important people during times of epidemic one went through the trial of fire and water at their doors. Washing with water and enveloping the body in smoke or incense was felt to be an effective defence against Black Death, and in addition torch bearers with plague torches (brands of burning fragrant herbs) walked ahead of important and rich personages.

Gentile of Foligno, a medical practitioner of Padua, made the connection between smell and death thus:

> *Poisonous material is generated about the heart and lungs. Its impression is not for excess in degree of primary quality, but through the properties of poisonous vapours having been communicated by means of air breathed in and out, great extention and transition of the Plague takes place, not only from man to man, but from country to country.*

The Plague doctors wore full, leather coats and a hat and crystal goggles to ensure they did not catch the disease from patients, and to be doubly sure touched them only with examining wands. They wore large cones, or nosegays, resembling a bird's beak or bill, and thus it is believed the derogative 'quack' applied to fake doctors.

In an effort to keep disease at bay, the first attempts to adopt more hygienic customs began around this time, at least with the aristocracy. The English King Henry III (1207–1272) sent the following message ahead during one trip to London:

> *Since the privy chamber in London is situated in an undue and improper place, wherefore it smells badly, we command you on the faith and love by which you are bounden to us that you in no wise omit to cause another privy chamber, to be made in such more fitting and proper place that you may select there, even though it should costs a hundred pounds . . .*

Guilds for the supply of aroma products began to be formed between the twelfth and thirteenth centuries. Related crafts included the London Guild of Pepperers and Spicers, and in 1268 the Glover's Guild was recognized. King Henry I of France and England granted a heraldic shield to the Guild of Perfumers, which was essentially silver, with three red gloves and a gold spice box on a blue background. The connection with gloves came from the need for leather tanners to hide the vile smell of their profession, and from their expertise in fine aromas the development of perfume grew. A charter to glover perfumes had been granted by Philip Augustus of France (1165–1223) as early as 1190.

Venice was an important centre for trade and commerce between Europe, the Middle East and the Orient, and became the funnel through which many spices and aromatic raw materials reached Europe, and its domination in trade for these products lasted for a few hundred years.

Glass saw major strides in its development. In medieval times a green tint, caused by iron in the silicate, was common. Antimony, used as a decolourizer, was replaced by manganese, and the Renaissance produced rapid development in the art of glassmaking in Venice. By 1400, a glass-like rock-crystal had been produced and perfected (*cristillo*).

Distillation as an art was well known in the eleventh century, but the first European treatise on distilling was written by the Catalonian Arnald of Villanova around 1310. Different types of distilled spirit were identified as *aqua vitae* (life), *aqua vini* (wine), and *aqua ardens* (burning water) and the book on the practice (*The Vertuose Boke of Distyllacyon*) was translated into English from the German 1500AD original version of Jerome Brunschweig by Lawrence Andrews. This book dealt in detail with the essential oils of lavender, juniperwood, pine and rosemary.

Early processes of distillation used alembics, usually made of copper, iron or tin, since lead and silver had the characteristic of tainting the distillation vapour.

Arnald of Villanova showed interest also in the sulfur baths of Montpellier, and it was around this time that the great fragrance raw material and production centre of Grasse, in the south of France, began to develop strongly. Along with Arnald of Villanova came such famous alchemical names as Roger Bacon (attributed with the invention in Europe of gunpowder), Ramon Lull of Spain, Nicolas Flamel of France and George Ripley of England, who kept the flames of knowledge alive, moving technique and thought apace. Table 2.1 gives a more detailed list of key figures in the history of alchemy, who furthered the development of an art which eventually became true chemistry. Alchemists, as we shall see, figured in the furtherance of aromatic knowledge for the next two centuries.

THE AGE OF DISCOVERY

The sixteenth and seventeenth centuries saw an explosion in world exploration. It had been preceeded by the Italian Colombus's voyages

Table 2.1 *Alchemists of history*

Date	*Provenance*	*Alchemist or related profession*	*Major work*
~1550BC	Egypt	Maria Prophetessa	Development of chemical apparatus
~1550BC	Egypt	Hermes Trismegitus	*The Emerald Tablet*
~384–322BC	Greece	Aristotle	*Secretum Secretorum*
~300–250BC	Egypt	Bolos of Mendes	*Phusika Kai Mustika*
~130AD	China	Wei Po Yang	*Ts' An T'ung Ch'i*
370–460	Greece	Democritus	Formulated early atomic theory
721–815	Persia	Jabir Ibn Hayyan	*Summa Perfectionis*
866–921	Persia	Rhazes	*Book of the Secret of Secrets*
980–1037	Persia	Avicenna (Ibn Sina)	*Book of The Remedy*
1193–1280	Germany	Albertus Magnus	*Libellus de Alchimia*
1214–94	England	Roger Bacon	*Opus Maius*
1235–1311	Spain	Arnald of Villanova	*A Treatise on the Preservation of Youth*
1235–94	Spain	Ramon Lull	*Dignitates Dei*
~1330	Italy	Petrus Bonus	*Pretiosa Margarita Novella*
1330–1417	France	Nicolas Flamel	*Work on Transmutary Alchemy*
~1450s	England	George Ripley	*Medulla Alchimiae*
~1470s	England	Thomas Norton	*The Ordinall of Alchimy*
1462–1516	Germany	Abbot Trithemius of Sponheim	*Steganographia* (Angel Magic)
1486–1535	Germany	Henry Agrippa	*Occulta Philosophia*
1493–1541	Germany	Paracelsus	*Der Grosseren Wundartzney*
1494–1555	Germany	Georgius Agricola	*De Metallica*
1527–1608	England	John Dee	*Monas Hieroglyphica*
1544–1609	France	Joseph Duchesne	*On the Material of the Medicine of Ancient Philosophers*
1548–1600	Italy	Giordano Bruno	*De Umbris Idearum*
1550–1627	England	Anthony Francis	*Panacea Aurea*
1568–1622	Germany	Michael Maier	*Atalanta*
1575–1624	Germany	Jacob Boehme	*On the Three Principles of Divine Being*
1586–1654	Germany	Johann Valentin Andrea	*Chemical Wedding of Christian Rosencreutz*
1560–1603	Germany	Heinrich Khunrath	*Amphitheatre of Eternal Wisdom*
1560–1616	Germany	Andreas Libavius	*Alchemia*
1574–1637	England	Robert Fludd	*Physics and Technics*
1580–1609	Germany	Oswald Croll	*Basilica Chemica*

Continued

Table 2.1 *Continued*

Date	*Provenance*	*Alchemist or related profession*	*Major work*
1603–1665	England	Sir Kenelm Digby	*A Choice Collection of Chymical Secrets*
1616–1654	England	Nicholas Culpeper	*A Physical Directory*
1617–1669	England	Elias Ashmole	*Theatrum Chemicum Britannicum*
~1620s	Hungary	Daniel Stolz	*Chemical Garden*
1627–1666	USA	George Starkey	*The Marrow of Alchemy*
1627–1691	England	Robert Boyle	*Sceptikal Chymist*
1642–1727	England	Sir Isaac Newton	*Principia*
born 1643	England	William Backhouse	*The Magister*

to the Americas, with John Cabot the Englishman close on his heels. Verrazzano searched for a Northern Strait (1524–1528), whilst Cartier too plied the coast of North America (1534–1536) in search of riches.

Queen Elizabeth's master mariners (Gilbert, Frobisher, Drake and Raleigh) trawled the seas in search of treasure for her coffers, whilst Magellan and da Gama probed the southern seas.

All of this marine activity brought back many new aroma products to the tables and laboratories of Europe, and increased the supplies of existing ones. And at the other end of the voyage the alchemists were waiting to add to their store of knowledge.

Court alchemists included Giordano Bruno (Henry III of France, 1551–1589), John Dee (Elizabeth I, Charles I) and Joseph Duchesne (Henry IV of France, 1553–1610).

Bruno was an early atomist, writing on *The Principles, Elements and Causes of Things* (1590), whilst Andreas Libavius in 1597 developed the study of alchemy in two directions, encheiria, the manipulation of materials, and chymia, the preparation and classification of chemicals. Biringuccio wrote of fireworks (1540, *Pirotechnia*), Agricola of metals (1556, *De Metallica*), Neri of glass manufacture (1612, *L'Arte Vetraria*), and Robert Fludd (1574–1637) covered nearly everything in his manifest works.

Meanwhile, Paracelsus (1493–1541) worked on distillation to separate the 'essential' from the 'non-essential' parts of a compound, and developed further the *quinta essentia* (quintessence) theory of a fifth element, involved in imbuing life. His radical ideas greatly influenced medicine during the Renaissance. According to Paracelsus, God makes medicine, but not in a prepared form. In nature, medicine is found compounded with 'dross', which must be taken away by distillation, setting the medicine free. The process became refined.

It was at the end of the seventeenth century that the German Johann Kunckel discovered how to use gold chloride to manufacture ruby glass, and also how to 'strike' red (a re-heating technique to develop the colour).

In 1573 Edward de Vere, Earl of Oxford, brought Elizabeth I not only scented sachets, but also perfumed gloves and jerkins. Entranced, Elizabeth, the Queen who had 'a bath every three months whether she needed it or no', became a fragrance enthusiast and the use of fragrances gradually became *de rigueur* in court.

Around this time the first books and manuscripts describing perfumery techniques surfaced, and court perfumers took the stage. A contemporary of Elizabeth, Catherine de Medici (1519–1589) travelled to France to marry Henry II, and in her entourage were two skilled artisans, Tombarelli and Renato Bianco, skilled in the crafts of perfumes and poisons, since court intrigue mixed affairs of the heart with affairs of the sword. Diane de Poitiers, a rival of Medici, was said to dabble herself in philtres, potions, perfumes and poisons. Nostradamus, the personal astrologer of Catherine, was known to inhale smoke and incense as part of his preparations for prophesying. As a plague doctor, Nostradamus used rose petal pills as a palliative and part herbal remedy for bad breath and cleaning teeth. The recipe for these pills included red roses picked before dawn, sawdust from fresh green cypress, iris, cloves, calamus, tiger lily and aloes. His second wife, Anne Ponsard Gemelle, was famed as a maker of herbalized perfumes.

Throughout the ages, perfume has provided a pathway to happiness; like happiness itself, the odours are intangible and often fleeting. History is littered with examples of the famous and their perfumed preferences: Henry III was said to have fallen head over heels in love with Mary of Cleeves after breathing the odour of her just removed clothing. Henry IV of France was reputed to smell so ripe that his intended, Marie de Medici (1573–1642) keeled over when she first met him, while Henry himself, revelling in his own natural odour and those of others, once reputedly wrote to his mistress Gabrielle d'Estree, 'Don't wash my love, I'll be home in eight days'. The French kings and their courts greatly indulged the use of fragrance, Louis XIII favouring neroli, based on orange blossom, whilst his chief adviser Cardinal Richelieu had the fragrant scent of flowers 'bellowed' through his apartments. Louis XIV, the Sun King, with his mistress Madame de Montespan, compounded his own fragrances, whilst

Louis XV lavished wealth on 'La Cour Parfumee' (the Perfumed Court) with his mistresses Madame de Pompadour and Madame du Barry, where even the fountains did not escape a fragrant dousing. Eventually the Madames of France, ending with Marie Antoinette, wife of Louis XVI (1774–1793) paid for their indulgence in meeting another Madame: the guillotine.

Meanwhile, over in England, Charles I (1600–1649) had Nell Gwynne as fragrant advisor, whilst Charles II (1630–1685) was encouraged in the aromatic arts by Catherine of Braganza. Perfume rings, filigree pomanders and vinaigrettes gave new ways to perfume the air, much needed since the strong smell of valerian musk and civet was more desirable to the lack of personal hygiene which existed at the time. The animalic link to humans had also been made, as instanced by this sage advice of a Mr Wecker in his *Secrets of Art and Nature*:

> *If any man would provoke a woman, let him sprinkle his Glans with Oyl, Musk or Civet, Castoreum or Cubeba, or any of these, for these so quickly provoke.*

THE AGE OF REVOLUTION

The rejection of monarchy, fuelled by an English Civil War earlier (1642–1651) grew apace in the revolutions of North America (1775–1783) and France (1785–1799), leading to fundamentally new politics and world order.

Meantime, a quieter revolution was occurring, as apothecaries, spicerers and chemists began to develop professionally a new trade, that of perfumery. In 1708, Charles Lilly, a London perfumer, introduced scented snuffs and a revolutionary fragrance consisting of orange flower, musk, civet, violet and amber, whilst in 1711 William Bayley opened a perfumery in Long Acre, moving later to Cockspur Street under the sign of 'Ye Olde Civet Cat'. Juan Floris (1730) and William Yardley (1770) added to the groundswell, matched in France by Houbigant (1774) and Lubin (1798).

These half-dozen perfumers catalysed the marketplace, expanding their clientele to commonfolk in such a way that George III (1738–1820) became concerned enough to issue the following edict:

> *All women whether of rank or professional degree, whether virgins, maids or widows, that shall from after this Act impose upon, seduce*

and betray into matrimony any of his majesty's subjects by the use of scents, potions, cosmetics, washes, artificial teeth, false hair, Spanish wool, iron stays, hoops, high heels, shoes or bolstered hips, shall incur the penalty of the law now in force against witchcraft and like misdemeanors, and that the marraige, upon conviction, shall be null and void.

Two famous compounded fragrances, *La Poudre de Marechale* (1670) and *Eau de Cologne* (1710), grew in popularity, marking a slight move in preference from the heavy animalic scents of the times, but with pox and pestilence to counter, aromas of all descriptions were slapped on, sprinkled over and carried in nosegays. Houses were refreshed using pomanders, potpourris, and cassoulets. The churches frowned and Oliver Cromwell did his best to put a stamp on the use of fragrances, but the eighteenth century saw a fragrance backlash of mighty proportions.

Fragrances were needed in profusion to combat the olfactory disaster zones of prisons, hospitals, ships, churches, theatres, workshops and, indeed, anywhere where there was a gathering of humming humanity. Major cities sported ooze-smelling rivers, malignant vapours and rank-smelling fogs from industry. Dyers, tanners, butchers and various artisans added to the smoke, coal, sulfur and fetid aromas in the air, and the environment was further fouled by the lax sanitary conditions of the time.

Balsams and light florals were in with a vengeance, with rose, violet and lavender leading the way. The Elizabethan 'effluvia' maskers were on their way out, and, after 300 years, not before their time. Like Casanova, European gentry had palled to the smells which evoked this comment from their distant ancestors:

Lady I would descend to kiss thy hand
But 'tis gloved, and civet makes me sick.

Meanwhile, the glass cutter borrowed techniques from the gem cutter, and the Venetian style of soda-lime glass was copied throughout Christendom, but without the same brilliance of metal. In 1673, the Glass Seller's Company of England commissioned George Ravenscroft to produce an acceptable substitute, and using up to 30% lead oxide in the mix, a brilliant glass with a high refractive index was manufactured, establishing England as a leader in the production of clear, decorative glass.

For colour, lithyalin glass, with slight metallic inclusions, gave hues ranging from reddish brown through leek green and olive green to bluish mauve, the better to hide impurities in a perfume mix. Hyalinth gave black glass, whilst selenium produced pink, oxides of cobalt and copper blue, cadmium sulfate, antimony and gold chloride yellow and the oxides of chromium and copper ruby glasses. Apart from hiding impurities, dark glass had a useful purpose in matching costumery and fashion at the time, and also in protecting a fragrant mix against ultraviolet light. At first, glass was seen as somewhat of a luxury, and thus its use focused around perfume, cosmetics and toiletries. Four main types of container evolved, the cylindrical or cigar-shaped alabastron, the pear-shaped amphoriskos, the short-necked, globe-like aryballos and the simple jug with a handle and flat base.

By the seventeenth century perfumes had begun to be stored in lightly blown glass bottles, and the eighteenth century saw the appearance of pear-shaped bottles in opaque white glass, decorated similarly to porcelain ware. Weight was reduced, and decorative appeal achieved by colour, cutting and appliqué decoration, which made perfume bottles truly treasured possessions, and worth much to today's collectors.

THE AGE OF EMPIRE

Napoleon Bonaparte loved aromas, even dispelling a revolution with his 'whiff of grapeshot'. Hailing from Corsica, Bonaparte liked the fresh citrus and herbal smells, and favoured *Eau de Cologne*, using by all accounts several bottles a day and more than 60 a month! In an echo back to the days of the French King Henry, he too extolled Josephine by messenger from his campaigns not to wash as he was returning home (the anecdotal accepted origin of the name for 'Je Reviens').

Josephine, a Creole from Martinique, employed a different potpourri of smells. She favoured animalics, and was particularly fond of musk oil, to the point that when Napoleon left her for another woman, she smeared the inside walls and curtainings of their house, Malmaison, with the all-pervading tenacious smell as a constant reminder of her. Needless to say, Napoleon was not too enamoured of that Parthian shot, but he still nevertheless kept a place in his heart for Josephine.

The old warrior carried a necklace with her cameo portrait and a pressed violet, one of her favourite flowers, around with him, and on her death visited her grave to strew it with violets. One of his own dying wishes was to be buried in a certain spot at St Helena where his soul would be soothed by the sweet smell of tuberose.

Queen Victoria can be credited with two 'smell' revolutions that hit Britain. The first involved her dress style, which often in casual mode featured a knitted shawl. The shawls she wore were steeped in patchouli, imbuing a rich woody fragrance to the garment. Since half the populace wished to emulate the monarch, patchouli shawls were *de rigueur*; this is one of the first instances of smell being used as a product plus in the marketing of a fashion item.

Victoria's second claim to fame was in marrying Prince Albert from the German house of Saxe-Coburg. In doing so that marvellous custom of the Christmas tree, long favoured in Central Europe and Scandinavia, was brought into British homes. The Christian festival of Christmas made much of the bringing of greenery into the home, as this signalled the return of springtime and the renewed growth of plant life on which we all depend. The Christmas tree brings an evocative smell of nature to life in our households, and along with clove, cinnamon, candle wax, mistletoe, holly and a cornucopia of scents, ensures that the memory of a merry Christmas is retained.

The wearing of perfume itself, however, was strictly controlled; just a little dab, or carried on a kerchief, never on one's person, was the rule of the day. Victorian ladies had to be 'proper' and scent was too evocative to be worn by a lady! As the Victorian era drew to a close, new names cropped up to cater for the mass-market demand in quality and reliability of scents. Scientists and artisans developed into perfumers of both integrity and repute. To famous names such as Lillie Yardley, Lentheric, and Floris were added Savoury and Moores, Atkinsons, Chardin, Crown, Coty, Hougibant, Guerlain, Roger and Gallet, Penhaligon and Piver; names which are familiar in households today.

Perfumery developed in three fundamental ways: the technique used, the structure and synthetics employed and the industrialization (massification) of the process.

Technique

In his book *Odours, Fragrances and Cosmetics* (1865), S. Piesse developed theories that related specific odours to notes on a musical scale in an attempt to categorize the spectrum of smells, whilst in 1890 Atkinsons produced one of the first books on perfume technology, essentially concerned with the production of absolutes by the cryoscopic removal of fats. Perfumery was beginning to be investigated in depth.

Structured Perfumes, and Use of Synthetics

In 1861, Guerlain created *Eau Imperiale* for Empress Eugenie, the influential wife of Napoleon III, whose gowns were designed by the House of Worth. By the end of the century, this redeveloped fragrance was shown to be created around neroli, rose, geranium, sandalwood, musk and the synthetic chemical coumarin. Fragrances began to be described in a structural form, with the adoption of top-, middle- and bottom-note terminology.

Two other fragrances, *Fougere Royale* (1882, Houbigant) and *Jicky* (1889, Coty) were in vogue. *Fougere Royale* was arguably amongst the first fragrances to use a synthetic (coumarin), whilst *Jicky* is held to be the first truly vertically structured fragrance, with a fresh, citrus top based on lemon, bergamot and mandarin, middle floral notes of rose and jasmin, woody notes in vetiver, orris root and patchouli, and base notes of coumarin, benzoin, civet, amber and vanillin (a second synthetic). *Fougere Royal* disappeared a long time ago, but its influence lives on in the aromatic fougere family of fragrance to which it lends its name.

Industrialization and 'Massification'

By 1879 it was listed that Yardley exported over a score of different varieties of scented soaps to the United States, whilst the British company Crown Fragrances was exporting 49 different fragrances to 47 different countries. Perfumers focused on mass production techniques for aroma chemicals, glass bottles and alcohol to service an ever-growing market demand. Products were branded to encourage consumer loyalty, and the first tentative steps to marketing their olfactory wonders were being made by the giants of their time. Perfumers of France, England and Spain widened their horizons to a global marketplace. At the end of the nineteenth century, science, industrialization, market demand and individuals of the moment had conspired to catalyse growth in the use of scent and the pleasure gained from it. Perfume was finally reaching the masses. This set the scene for the twentieth century, the age of fashion, which spurred a truly explosive growth in the use of fragrance in many forms. In the background of this flurry of activity on the perfume front, major strides had been taken in the synthesis of aroma chemicals which greatly influenced and aided the formulation chemists and perfumers in their choice and cost of materials. Table 2.2 summarizes some of the key

Table 2.2 *Important Dates in the History of Aroma Chemicals*

Year	*Event*	*Attributed to*
1701	Observations that some flowers provided no essential oils on steam distillation	Nicholas Lemery
1759	Reaction of oil of amber with fuming nitric acid gave a musky odour	Berlin Academy
1800	Investigations into ambra (or ambergris) component chemistry	
1833	Empirical formulae reported for anethole, borneol and camphor	Dumas
1834	Isolation of cinnamic aldehyde	Dumas, Peligot
	Preparation of nitrobenzene	Mitscherlich
1837	Isolation of benzaldehyde	Liebig, Wohler
1843	Methyl salicylate determined as main component of wintergreen oil	Cahours
1853	Preparation of benzyl alcohol	Cannizzaro
	Synthesis of aliphatic aldehydes	Piria
1856	Synthesis of cinnamic aldehyde	Chiozza
1859	Preparation of aldehydes from pyrolysis of calcium formate mixtures	Bertagnini
1859–1860	Large-scale preparation of salicylic acid	Kolbe
1863	Preparation of benzaldehyde	Cahours
1865	Determination of structure of benzene	Kekule
1866	Structure of cinnamic acid determined	Erlenmeyer
1868	Synthesis of coumarin	Perkin
1869	Discovery of heliotropin	Filtig and Mielk
1871	Structure of heliotropin determined	Barth
1874	Synthesis of vanillin from guaiacol	Reimer and Tiemann
1876	Discovery of phenylacetic aldehyde	Radziszewski
1875–1877	Synthesis of cinnamic acid	Perkin
1877	Production of anisaldehyde from *p*-hydroxybenzaldehyde	Tiemann, Herzfeld
1878	Structure of terpin hydrate determined	Tilden
1880	Quinolines discovered	Skraup
1884	Identification of D-limonene and dipentene	Wallach[a]
1888	Discovery of nitro musks	Baur
1889	Discovery of citronellal	Dodge
1890	Synthesis of heliotropin from safrole	Eykmann
1893	Synthesis of ionone	Tiemann, Kruger
1894	Structure of α-pinene determined	Wagner
1885	Structure of terpineol determined	Wallach, Tiemann, Semler

Continued

Table 2.2 *Continued*

Year	*Event*	*Attributed to*
1891	Discovery of rhodinol	Eckhart
1898	Discovery of musk ketone	Baur, Thurgau
1903	Discovery of methyl heptin carbonate and homologues	Moureau, Delange
	Fundamental work on aromaticity	Von Baeyer[a]
1904	Synthesis of methylnonylacetic aldehyde	Darzens
	Isolation of muscone	Walbaum
	Glycidic method of synthesizing aldehydes	Darzens
1905	Synthesis of cinnamic alcohol	Leser, Barbier
1905–1908	Hydroxycitronellal prepared and marketed	Knoll and Co.
1908	Discovery of γ-undecalactone	Jukov, Schestakow
1913	Discovery of farnesol	Kerschbaum
1919	Discovery of cyclamen aldehyde	Blanc
	Synthesis of linalool	Ruzicka, Fomasir
1923	Discovery of α-amyl cinnamaldehyde	Lesech, Descollonges
	Discovery of Nerolidol	Ruzicka
1926	Identification of muscone structure	Ruzicka[a] and
	Structural determination work on ambra, civet	Kerschbaum
	Synthesis of Exaltone®	Ruzicka
1927	Isolation of Ambrettolide®	Kerschbaum
	Synthesis of Civetone	Ruzicka
	Synthesis of Exaltolide®	Kerschbaum
1933	Isolation of Jasmone	Ruzicka
1934	Synthesis of muscone	Weber, Ziegler
1935	Structural determination of jasmone	Treff, Werner[a]
1946	Perfection of Wallach's isoprene rule for terpinoids	Robinson[a]
1947	Structural determination of irone	Ruzicka,[a] Naves
1950	Synthesis of Ambrox®	Stoll
1953	New synthesis of linalool	Caroll, Kimel
1960	Synthesis of *cis*-Hexene-3-ol	Hakanata
1962	Synthesis of Methyl dihydrojasmonate	Firmenich
1970	Synthesis of Damascenones alpha and beta	Demole
1971	Synthesis of Damascones alpha and beta	Ohloff

[a] Five Nobel prize winners in the first half of the twentieth century were involved with aspects of aroma chemistry.

Table 2.3 *Twentieth Century: The Age of Fashion*

Decade	*Background*	*Development and Exploitation*	*Fragrance used in*	*Emergent fashion designers*
1900s	*Fin de Siècle*, emancipation	Coumarin, heliotropin, ambreine	*L'Origan*	Worth
1910s	Peace and war	Undecalactone	*Mitsouko*	Poiret, Caron
1920s	Prohibition, exhibition	Aldehydes C10, C11, C12	*Chanel N°5*	Chanel, Patou
1930s	Recession, depression	Phenyl ethyl acetate, civettone	*Tabu*	Schiaparelli, Dana
1940s	War and peace	Hydroxycitronellal, musk ketone	*L'Air du Temps*	Dior, Balenciaga, Balmain
1950s	Rock and roll	Amyl salicylate, cedryl acetate, nitromusks	*Youth Dew*	Rochas, Nina Ricci
1960s	Flower power	PTBCHA,[a] *cis*-hex-3-salicylate	*Fidji*	Laroche, Quant
1970s	Global village	Methyl dihydrojasminate	*Chanel N°19*	Cacharel, Paco Rabanne
1980s	King consumer	Ethylene brassylate, helional	*Obsession*	Montana, Jean Paul Galtier
1990s	*Fin de Siècle*, millennium	Dihydromyrcenol, synthetic musks, Ambrox© (Firmenich)	*Cool Water*	Thierry Mugler, Hugo Boss, Joop

[a] *p*-*t*-Butylcyclohexyl acetate.

compounds of interest to the industry, their discovery dates and the chemists involved.

THE AGE OF FASHION

I am no longer interested in dressing a few hundred women, private clients; I shall dress thousands of women. But ... a widely repeated fashion, seen everywhere, cheaply produced, must start from luxury.
Gabrielle (Coco) Chanel

Table 2.3 underscores a prodigious growth in the use of fragrances, where for each decade of the twentieth century, against dramatically different social backdrops, novel chemistry was developed that gave new strength, depth and vision to the world of perfume. However, it took another phenomenon to catalyse the fine fragrance industry to the level that we see today: the fashion designer and the rise of the consumer.

Whilst the perfume companies brought the baton of perfume into the twentieth century, and still run in the games, it was the designers who took the baton and ran a different type of race, a race to bring a name to the masses.

In the first 20 years of the twentieth century, a score of fine fragrances was developed, including *Violette Purpre* (1907, Houbigant), *L'Origan* (1905, Coty), *English Lavender* (1910, Atkinsons), *L'Heure Blue* (1912, Coty), and *Old English Lavender* (1913, Yardley). During the last decade of that century the industry had grown to such an extent that over 100 fine fragrances a year were being launched. Perfume had finally come to the people. Chemistry and creativity had brought it there.

In 1905, Francois Coty said, 'Give a women the best product you can compound, present it in a container of simple, but impeccable taste, charge a reasonable price for it, and a great business will arise such as the world has never seen.'

The man was not only a genius, but a visionary as well.

Chapter 3

Perfumery Materials of Natural Origin

CHARLES SELL

PERFUMES AND ODOURS IN NATURE

Introduction

Like the pharmaceutical industry, the fragrance industry uses nature as its guide and source of inspiration. All the perfumes and perfume ingredients that we produce in our factories are modelled to a greater or lesser extent on those found in nature. We observe nature, analyse it to find out how it does the job and then modify and adapt its methods to suit our needs.

Smell and taste are the oldest of our senses. They probably developed in very primitive organisms as a means of obtaining information about chemical changes in the organism's environment. Diurnal birds and aquatic animals rely heavily on sound; man and a few primates rely on vision; but all other species use smell and taste, the chemical senses, as the dominant medium through which they obtain information about the world in which they live. Since smell is such an important source of information for us, it is not surprising that nature has developed a very sensitive and sophisticated system for the analysis of the chemicals which make up our environment. It is intriguing that we can detect not only the natural odours, whether they have been placed deliberately or are simply artefacts of degradation processes, but also chemicals to which we have not been exposed before. We have only a very limited understanding of how this wonderful sense works; to find the solution to this riddle is one of the great scientific challenges of the moment.

Animals use smell and taste to find food and to assess its quality. The smell of food has a powerful effect on animals, whether it is a lion smelling out a herd of wildebeest or a shopper being drawn to the in-store bakery at the back of the supermarket. Watch your cat when you give it a bowl of food. It carefully sniffs the food before eating, to check that it is fresh and good. Our aversion to the smell of amines and mercaptans is, doubtless, related to their presence in food that has been spoiled by bacterial decomposition, a strong warning signal against sources of food poisoning. Some degradation reactions are responsible for the development of flavour in food. For example, autoxidation of fatty acids can lead to the formation of materials of characteristic aldehydic flavour. It is important to note that the tongue (*i.e.* the sense of taste) only detects sweet, salt, sour and bitter. The rest of taste is, in fact, smell. The volatile flavour ingredients are vaporized in the mouth and reach the nose through the airways behind the roof of the mouth.

Living organisms also use the chemical senses as a means of communication. If the communication is between different parts of the same organism, the messenger is referred to as a hormone. Chemicals used to carry signals from one organism to another are known as semiochemicals, which can be grouped into two main classes: pheromones and allelochemicals. If the signal is between two members of the same species, the messenger is called a pheromone. Pheromones carry a variety of types of information. Not all species use pheromones, but in those that do, some may use only one or two pheromones whilst others, in particular the social insects such as bees, ants and termites, use an array of chemical signals to organize most aspects of their lives. Sex pheromones are amongst the most widespread. Male moths can detect females by smell at a range of many miles. Androstenol is the compound that produces 'boar taint' in pork. It is produced by boars and is released in a fine aerosol when the boar salivates and champs his jaws. When the sow detects the pheromone in the air, she immediately adopts what is known as 'the mating stance' in readiness for the boar. Ants and termites use trail pheromones to identify a path between the nest and a food source, which explains why ants often walk in single file over quite long distances. The social insects also use alarm, aggregation, dispersal and social pheromones to warn of danger and to control group behaviour. Chemicals that carry messages between members of different species are known as allelochemicals. Within this group, kairomones benefit the receiver of the signal, allomones its sender and with synomones both the sender and receiver benefit. Thus, the scent of a flower is a synomone since the attracted insect finds nectar and the plant obtains a pollinator. Some plants produce compounds known as antifeedants, the taste of which insects find repulsive. These are allomones since the signal generator, the plant, receives the benefit of not being eaten.

Unlike pheromones, many odorous chemicals in nature are produced for properties other than their odour. Many plants, when damaged, exude resinous materials as a defence mechanism. The shrub *Commiphora abyssinica*, for example, produces a resin that contains a number of antibacterial and antifungal compounds. The role of the resin is to seal the wound and prevent bacteria and fungi from entering and damaging the plant. The resin has a pleasant odour and so was put to use by man as a perfume ingredient. It is known as myrrh. As a result of its antimicrobial properties, myrrh was also used as an antiseptic and preservative material, for instance, in the embalming of corpses. Frankincense has been used in religious rites for thousands of years, and so two of the three gifts brought to the Christ Child by the Magi were perfume ingredients. Knowledge of perfumery thus helps us to understand the symbolism involved; gold, frankincense and myrrh represent, respectively, king, priest and sacrifice.

Biosynthesis

So, plants and animals produce odorous materials for a wide variety of reasons, but how do they generate them? All living organisms produce chemicals through a process known as biosynthesis. The materials thus produced can be classified into two major groups, *viz.* primary and secondary metabolites. Primary metabolites are those that are common to all species and can be subdivided into proteins, carbohydrates, lipids and nucleic acids. The materials used as perfume ingredients are mostly secondary metabolites, though a few are derived from primary metabolites by degradative processes. The four categories of secondary metabolites, in decreasing order of importance as sources of perfume ingredients, are terpenoids, shikimic acid derivatives, polyketides and alkaloids. Very few odorous materials are derived from the alkaloid family, so these are not discussed further here. Of the others, the terpenes are, by far, the most important. The terpenoids, shikimates and polyketides are all originally derived from glucose (Scheme 3.1; in this scheme and subsequent ones, the letter P is used to represent a single phosphate unit). It is worthwhile spending some time considering how the natural perfume ingredients are put together since, through this, the patterns of nature can be understood and used to assist in identifying the structures of newly isolated materials and in producing new compounds with similar odour properties. More detail on biogenesis is given in the books by Bu'Lock and Mann *et al.* (1994).

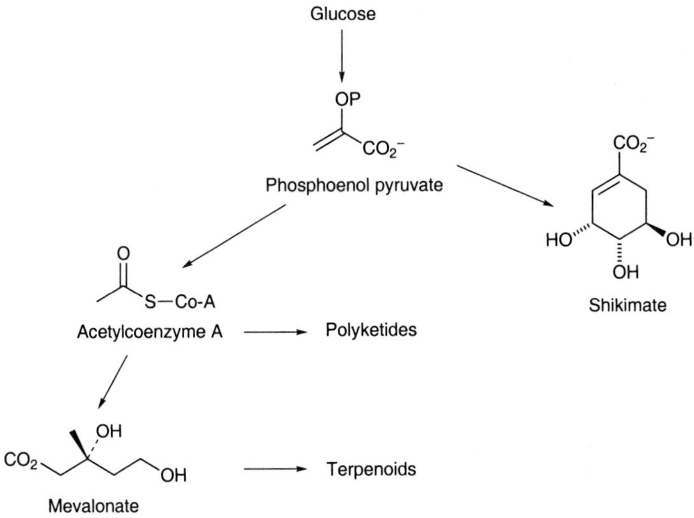

Glucose → Phosphoenol pyruvate → Shikimate; Phosphoenol pyruvate → Acetylcoenzyme A → Polyketides; Acetylcoenzyme A → Mevalonate → Terpenoids

Scheme 3.1

Green plants and photosynthetic algae synthesize glucose from carbon dioxide and water using sunlight as the energy source to drive this energetically unfavourable process, which is known as photosynthesis. Glucose can be broken down, either by the plant which made it or by another species which obtains it by eating the plant, to give the enol form of pyruvic acid, in which the enolic hydroxyl group is protected by formation of a phosphate ester. One metabolic pathway builds shikimic acid from the phosphoenol pyruvate and another converts it into acetyl coenzyme-A. The thiol function of coenzyme-A serves both as an activating group and as an efficient leaving group, thus making aldol-type chemistry facile and leading to long-chain compounds in which every second carbon existed, at some point, as a ketone. Self-condensation of these chains leads to the polyketides. Acetyl coenzyme-A can also be used to synthesize mevalonic acid, precursor to the terpenoids.

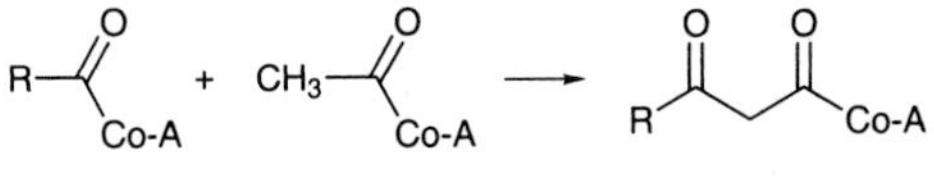

$$R{-}CO{-}Co\text{-}A + CH_3{-}CO{-}Co\text{-}A \longrightarrow R{-}CO{-}CH_2{-}CO{-}Co\text{-}A$$

Scheme 3.2

Lipids and polyketides are biosynthesized by aldol-type reactions of esters with coenzyme-A, as shown in Scheme 3.2. The coenzyme-A ester of a fatty acid undergoes reaction with acetyl coenzyme-A to give a β-ketoester. Reduction of the ketone group followed by elimination of the resultant alcohol and addition of hydrogen gives an acid with two carbon atoms more in the chain. This is why natural fatty acids contain even numbers of carbon atoms in their chains. If the polyketoacids undergo condensation reactions rather than reduction, the result is a phenolic material of the polyketide family, as in the formation of orsellinic acid (Scheme 3.3), which is the precursor for some odorous components of plants.

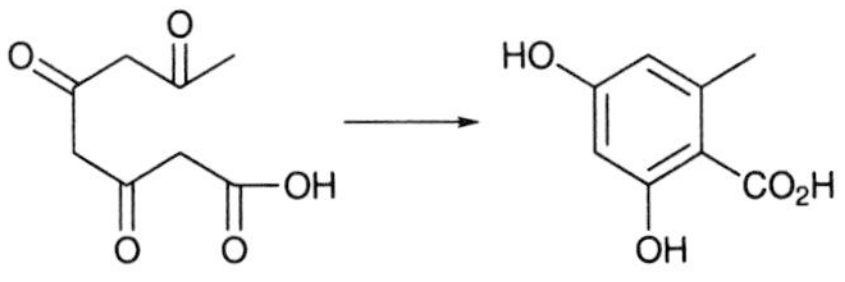

Scheme 3.3

One lipid of interest is arachidonic acid. This polyunsaturated fatty acid undergoes a radical cyclization reaction involving oxygen, as shown in Scheme 3.4. This cyclization leads to an important group of compounds known as prostaglandins, hormones in the animal kingdom. Degradative reactions lead to shortening of the chains to give jasmonic acid, a plant hormone and precursor for two important odorous materials, jasmone and methyl jasmonate.

Addition of phosphoenol pyruvate to erythrose-4-phosphate leads, through a number of reaction steps, to shikimic acid. The 3,4,5-trihydroxybenzoic skeleton of shikimic acid occurs in many perfume components, although the oxygen atoms in the product are not usually those of the original shikimic acid. The original oxygen atoms are lost during biosynthesis and others reintroduced into the same sites at a later stage by oxidation. Addition of a further unit of phosphoenol pyruvate adds a three-carbon chain to the carbon carrying the carboxyl group. The latter is then lost by decarboxylation. An abridged scheme for the biosynthesis of eugenol, the characteristic odorant of cloves, from shikimic acid is shown in Scheme 3.5.

Terpenes are defined as materials made up of isoprene (2-methylbutadiene) units. In the perfume industry the word 'terpene' is often used incorrectly to refer to monoterpene hydrocarbons. However, the term does include all compounds derived from the connection of two isoprene units to give a 10-carbon skeleton. The names given to the other members of the terpene family are shown in Table 3.1

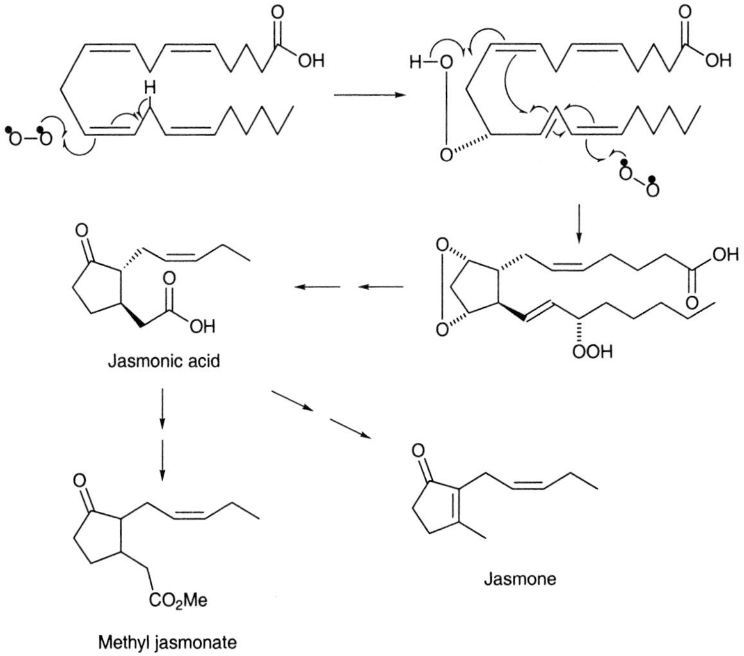
O
OH
H
H–O
O
O
OH
O
OOH
O
OH
Jasmonic acid
O
CO2Me
Methyl jasmonate
O
Jasmone

Scheme 3.4

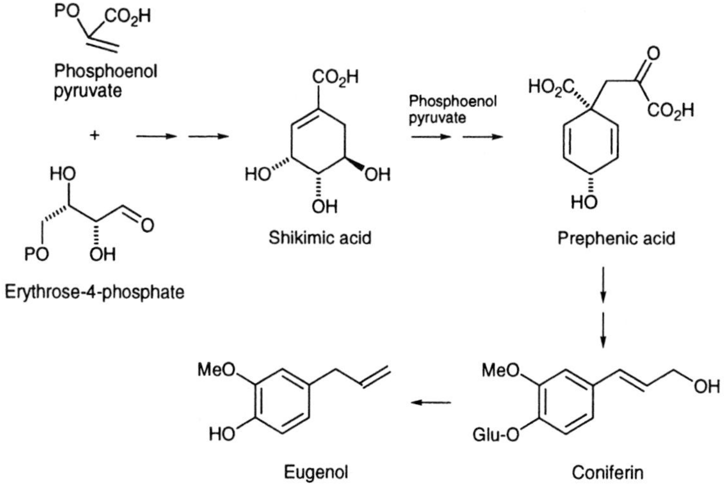
PO
CO2H
Phosphoenol pyruvate
+
HO
PO
OH
O
Erythrose-4-phosphate
CO2H
HO
OH
OH
Shikimic acid
Phosphoenol pyruvate
HO2C
O
CO2H
HO
Prephenic acid
MeO
HO
Eugenol
MeO
Glu-O
OH
Coniferin

Scheme 3.5

Table 3.1 *Classification of terpenes*

Name	*Number of isoprene units*	*Number of carbon atoms*
Hemiterpenes	1	5
Monoterpenes	2	10
Sesquiterpenes	3	15
Diterpenes	4	20
Sesterterpenes	5	25
Triterpenes	6	30
Carotenes	8	40
Steroids	Terpenoids which produce Diels's hydrocarbon when distilled from zinc dust	

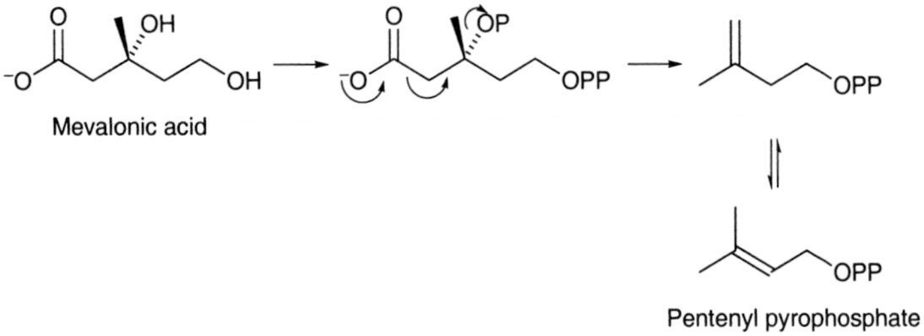

O OH $^-$O OH → O OP $^-$O OPP → OPP

Mevalonic acid

⇅

OPP

Pentenyl pyrophosphate

Scheme 3.6

Scheme 3.6 illustrates how, through phosphorylation, elimination and decarboxylation, mevalonic acid is converted into isopentenyl pyrophosphate, which can be isomerized enzymically into pentenyl pyrophosphate. Coupling of these two isomeric materials gives geranyl pyrophosphate, as shown in Scheme 3.7. Addition of a further molecule of isopentenyl pyrophosphate gives farnesyl pyrophosphate. These coupled units then lead to the monoterpenes and sesquiterpenes, respectively. Addition of further units of isopentenyl pyrophosphate leads in the same manner, to the higher terpenes. The reactions shown in Scheme 3.7 give rise to what is referred to as the head-to-tail coupling, in which the 'head' of one isoprene unit is connected to the 'tail' of another. This is, by far, the most common way of joining isoprene units together, though tail-to-tail couplings also occur, the best example being the tail-to-tail fusion of two geranylgeranyl pyrophosphate units to produce squalene and the carotenes. The terpene pyrophosphates undergo cyclization reactions under the influence of appropriate enzymes. Other enzymes then carry out further chemical conversions, such as oxidation, on the terpenes. This leads to a vast array of complex structures, the final structure depending on the exact nature of the enzymic reactions involved. Since the enzymes are often unique to one species, the terpenes (and of course, other metabolites also) produced by a plant can be used by botanists to classify it. Such classification of plants is referred to as chemotaxonomy.

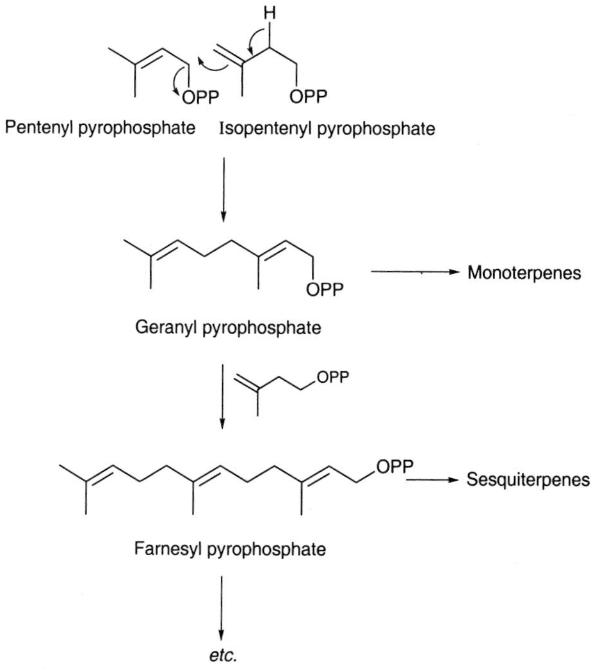

H
OPP OPP
Pentenyl pyrophosphate Isopentenyl pyrophosphate

OPP → Monoterpenes
Geranyl pyrophosphate

OPP

OPP → Sesquiterpenes

Farnesyl pyrophosphate

etc.

Scheme 3.7

Scheme 3.8 shows how the isoprene units and the original backbone can be traced out in a number of terpenes that are important in perfumery. Sometimes skeletal rearrangements occur which make this process more difficult and fragmentation or degradation reactions can reduce the number of carbon atoms so that the empirical formula does not contain a simple multiple of five carbons. Nonetheless, the natural product chemist quickly recognizes the characteristic terpene framework of the structure.

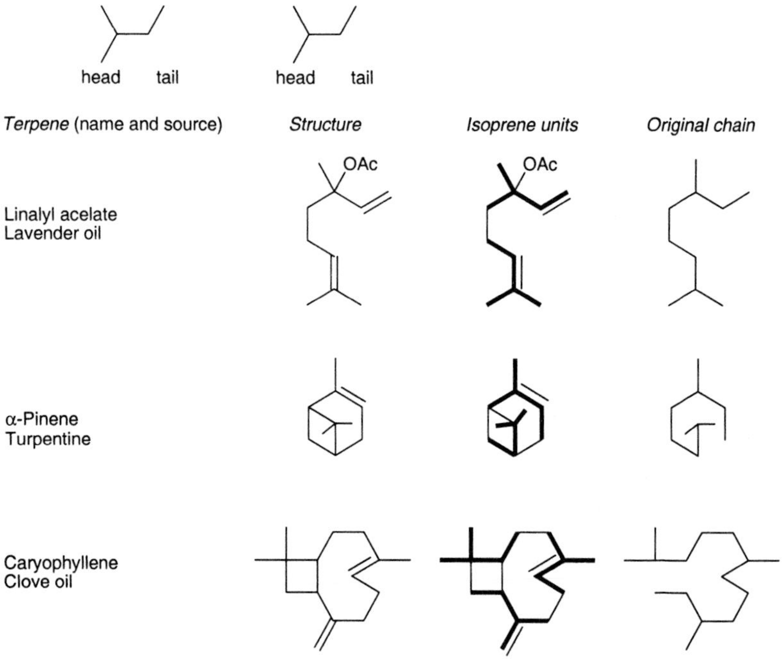

head tail head tail

Terpene (name and source) *Structure* *Isoprene units* *Original chain*

OAc OAc

Linalyl acelate
Lavender oil

α-Pinene
Turpentine

Caryophyllene
Clove oil

Scheme 3.8

The book by Mann *et al.* (1994) on natural products provides a good introduction to the biogenesis of natural perfume ingredients and the review by Croteau (1987) gives further detail on the biosynthesis of monoterpenes.

EXTRACTION OF NATURAL PERFUME INGREDIENTS

The methods used to extract perfume ingredients from their natural sources have changed over time as technology in general has advanced. However, both old and new methods fall into three basic classes: expression, distillation and solvent extraction.

Expression

Expression is the simplest of the three techniques. When odorants are forced out of the natural source by physical pressure, the process is referred to as expression and the product is called an expressed oil. If a piece of orange peel is squeezed, the oil bearing glands burst and eject a fine spray of orange oil. Many commercially available citrus oils are prepared in this way.

Distillation

Distillation of perfume ingredients from their natural sources can be done in three ways: dry (or empyreumatic) distillation, steam distillation or hydrodiffusion. Dry distillation involves high temperatures, since heat (and in most cases this is direct flame) is applied to the surface of the vessel containing the plant material. Usually this technique is reserved for the oils of highest boiling point, typically those derived from wood, because the high temperatures are necessary to vaporize their chemical components. Cade and birch tar are the major oils obtained by dry distillation. Cade and birch tar oils contain distinctive, burnt, smoky notes as a result of pyrolysis of plant material. In steam distillation, water or steam is added to the still pot and the oils are co-distilled with the steam. The oil is separated from the water by means of a Florentine flask, which separates them based on their differing densities. Figure 3.1 shows a simple schematic representation of a still and a Florentine and Figure 3.2 shows a still charged with jasmine flowers ready for the top to be fitted prior to distillation.

The waters that co-distil with the oil are called the waters of cohobation. In most cases, these are a waste product and are either

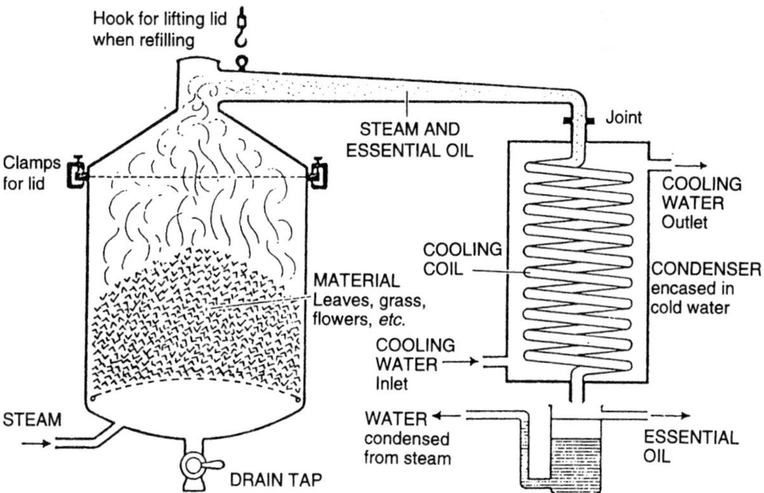

Figure 3.1 *Still and Florentine flask*

Figure 3.2 *Still charged with jasmine flowers*

discarded or recycled to the still pot. The waters of cohobation obtained from rose distillation are different. Rose oil is somewhat water soluble and so the 'rose water' is kept as a perfume and flavour ingredient. The presence of water in the pot during steam distillation limits the temperature of the process to 100 °C. This means that much less degradation occurs in this process than in dry distillation. However, some degradation does occur. For example, tertiary alcohols present in the plant often dehydrate in the pot and distil as the corresponding hydrocarbons.

Hydrodiffusion is a relatively new technique, and is essentially a form of steam distillation. However, it is steam distillation carried out upside down since the steam is introduced at the top of the pot and the water and oil taken off as liquids at the bottom.

Perfume materials obtained in this way are referred to as essential oils. Thus, for example, the oil obtained by steam distillation of lavender is known as the essential oil of lavender, or lavender oil. Sometimes, the monoterpene hydrocarbons are removed from the oils by distillation or solvent extraction to give a finer odour in the product. The process is known as deterpenation and the product is referred to as a terpeneless oil. This is, of course, a misnomer since, for example, the major component of lavender oil terpeneless is linalyl acetate, a monoterpene.

Solvent Extraction

Scheme 3.9 summarizes the various possible processes using solvent extraction to obtain perfume ingredients. The processes are written in lower case and the technical names for the various products in capitals.

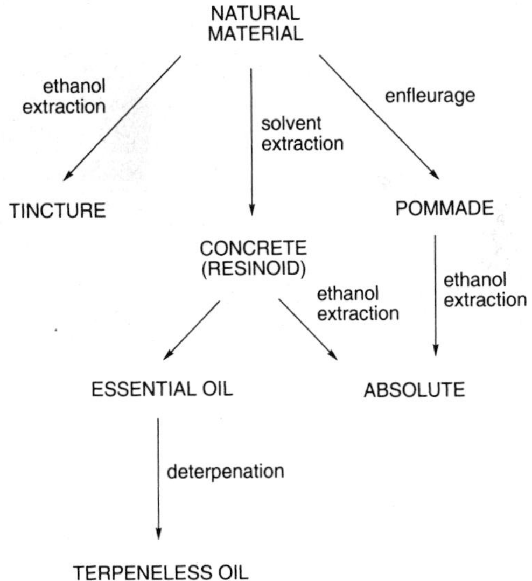

Scheme 3.9

Ethanolic extraction is not used very much for plant materials because of the high proportion of water compared with oil in the plant (vanilla beans are an important exception). It is more important with materials such as ambergris. The sperm whale produces a triterpene known as ambreine in its intestinal tract. This is excreted into the sea and, on exposure to salt water, air and sunlight, undergoes a complex series of degradative reactions which produce the material known as ambergris. (More detail of this chemistry is given in Chapter 4.) This waxy substance can be found floating in the sea or washed up on beaches. Extraction of it with ethanol produces tincture of ambergris.

Enfleurage was used by the ancient Egyptians to extract perfume ingredients from plant material and exudates. Its use continued up to the twentieth century, but it is now of no commercial significance. In enfleurage, the natural material is brought into intimate contact with purified fat. For flowers, for example, the petals are pressed into a thin bed of fat. The perfume oils diffuse into the fat over time and then the fat can be melted and the whole mixture filtered to remove solid matter. On cooling, the fat forms a pommade. Although the pommade contains the odorous principles of the plant, this is not a very convenient form in which to have them. The concentration is relatively low and the fat is not the easiest or most pleasant material to handle, besides which it eventually turns rancid. The ancient Egyptians used to apply the pommade directly to their heads, but in more recent times it became usual to extract the fat with ethanol. The odorous oils are soluble in alcohol because of their degree of oxygenation. The fat used in the extraction and any fats and waxes extracted from the plant along with the oil are insoluble in ethanol and so are separated from the oil. Removal of the ethanol by distillation produces what is known as an absolute.

The most important extraction technique nowadays is simple solvent extraction. The traditional solvent for extraction was benzene, but this has been superseded by other solvents because of concern over the possible toxic effects of benzene on those working with it. Petroleum ether, acetone, hexane and ethyl acetate, together with various combinations of these, are typical solvents used for extraction. Recently, there has been a great deal of interest in the use of carbon dioxide as an extraction solvent. The process is normally referred to as super-critical carbon dioxide extraction but, in fact, the pressures employed are usually below the critical pressure and the extraction medium is sub-critical, liquid carbon dioxide. The pressure required to liquefy carbon dioxide at ambient temperature is still considerable and thus the necessary equipment is expensive. This is reflected in the cost of the oils produced, but carbon dioxide has the advantage that it is easily removed and there are no concerns about residual solvent levels.

The product of such extractions is called a concrete or resinoid. It can be extracted with ethanol to yield an absolute, or distilled to give an essential oil. The oil can then be deterpenated. As noted earlier, the use of the word terpene here is misleading to the chemist since, in this instance, it refers specifically to monoterpene hydrocarbons. Hence, a terpeneless oil is one from which the hydrocarbons have been removed to leave only the oxygenated species and so increase the strength of its odour. With some particularly viscous concretes, such as those from treemoss or oakmoss, it is more usual to dissolve the concrete in a high boiling solvent, such as bis-2-ethylhexyl phthalate, and then co-distil the product with this solvent.

Essential oils and other extracts vary considerably in price and in the volume used each year. Lavender, for example, is a relatively inexpensive oil, costing £15–20/kg and 250–300 tonnes are used annually. Rose and jasmine are much more expensive and are used in much smaller quantities. The total annual production of rose oil is 15–20 tonnes and it costs between £1000 and £3000/kg, depending on quality. About 12 tonnes of jasmine extracts are produced annually at prices up to £2000/kg. Eucalyptus oil (from *Eucalyptus globulus*) has one of the largest production volumes, almost 2000 tonnes/annum and is one of the cheapest oils at £2–3/kg. The exact balance between volume and price depends on various factors such as ease of cultivation, ease of extraction and usefulness. For example, eucalyptus trees grow well, the leaves are easy to harvest, trimmed trees grow back vigorously, the oil is easily distilled and it is useful as a disinfectant as well as a camphoraceous fragrance ingredient. All of these factors combine to make it a high-tonnage oil.

Before this century, perfumes commanded such a price that only the wealthiest people could afford them. This is because perfumers relied on natural sources for their ingredients. Most of these ingredients are in limited supply and are expensive to produce. For instance, it takes about 7 000 000 jasmine flowers to produce 1 kg of oil. The flowers have to be picked by hand (no-one has yet devised a mechanical method of harvesting jasmine) in the first few hours of the day when their oil content is at its highest (Figure 3.3). In view of the costs of cultivation and extraction, it is not surprising to find that jasmine oils cost in the region of £2000/kg.

Some natural oils are much less expensive because of automated farming methods. For instance, rows of lavender in a field (Figure 3.4) can be cut almost to ground level and fed directly into a still pot carried on the tractor. The pot is then fitted under a field still and the oil extracted while the harvesting continues. The cost of lavender oil is thus tens, rather than thousands, of pounds *per* kilogramme. Despite this,

Figure 3.3 *Hand picking of jasmine*

Figure 3.4 *Cultivation of lavender*

the modern perfumery industry would be unable to function as it does if it were to rely solely on natural ingredients. Cost alone would be prohibitive, regardless of problems of stability in products or availability in view of limits on land use, *etc.*

Since essential oils are usually present in the botanical source at the level of only a percent or two, at most, of the dry weight of the harvested plant, it is more economic to extract the oil at the location where the plant grows, and ship the oil rather than the plant material, to the customer. The degree of sophistication of the harvesting and extraction technology varies widely, depending on the country of origin. The mint production of the USA and the lavender production of Tasmania are highly automated; indeed, they must be to remain economically feasible in countries with such high labour costs. In some other countries, simple bush stills constructed from waste oil drums and drainpipes are the most cost-effective means of production. Table 3.2 lists some of the more important of the essential oils used in perfumery today, and includes information on the plant parts and extraction techniques used to produce the fragrance products and also some of the more important countries of origin for each.

ADULTERATION OF NATURAL PERFUME INGREDIENTS

The high prices that essential oils command lead, inevitably, to the temptation for less scrupulous producers and dealers to adulterate the product. Adulteration is sometimes referred to, euphemistically, as sophistication. By adding lower cost materials, but still asking the same high price for the mixture, the person perpetrating the fraud can stand to make considerable sums of money from an unsuspecting buyer. However, the major fragrance companies are sufficiently astute and technically competent to uncover almost all attempts at such fraud. The techniques used in adulteration vary from the crude to the very sophisticated, as evidenced by the following examples. The examples also illustrate some of the quality control (QC) techniques used routinely by the industry.

A supplier of ylang-ylang oil once tried to sell drums that contained only a small amount of the oil, the remainder of the drum contents being river water. This attempt at deceit was easily discovered, since QC samples are normally drawn from top, middle and bottom of drums, so one sample would be oil and the other two water.

Lavender oils are relatively inexpensive essential oils, but their major components are available as even cheaper chemicals and so the possibility to cheat does exist. Like all natural products, the composition

Table 3.2 *Some of the more important natural fragrance materials*

Oil	*Types of process used*[a]	*Plant part extracted*	*Approximate annual production* (tonnes)	*Typical country of origin*
Ambrette	S	seed	0.5	China, Colombia
Angelica	S	root	1	Bulgaria
Anise	S	seed	1200	China, Vietnam
Artemisia	S	aerial parts	16	Morocco, Tunisia, India
Basil	S	flowering tops	15	Reunion
Bay	S	leaf	20	Dominca, Puerto Rico
Bergamot	E	fruit	120	Italy
Benzoin	C	exudate	3	Thailand, Indonesia
Birch tar	D	wood	50	Austria, Germany, Russia
Cabrueva	S	wood	10	Brazil, Paraguay
Cade	D	wood[b]	12	Portugal, Yugoslavia
Cajeput	S	leaves & twigs	50	Indonesia
Calamus	S	rhizome	10	N. Korea, India
Camphor	S	wood	250	China
Cananga	S	flowers	45	Indonesia, Comoros Islands, Reunion
Caraway	S	seeds	10	Bulgaria, Egypt, Australia
Cassia	S	leaves	160	China
Cedarwood[c]	S	wood	2200	China, USA
Cedar leaf	S	leaf	25	USA, Canada
Celery	S	seed	25	Bulgaria, India
Chamomile	S	flowers	10	Morocco, France
Cinnamon bark	S	bark	5	Sri Lanka
Cinnamon leaf	S	leaf	100	Sri Lanka, India, Seychelles
Citronella	S	leaves	2300	Sri Lanka, Indonesia
Clary sage	A, S	flowers/leaves	45	Russia, USA, Bulgaria, France
Clove bud	S	flower bud	70	Indonesia, Madagascar
Clove leaf	S	leaf	2000	Indonesia, Madagascar
Copaiba balsam	U	exudate	40	Brazil
Coriander	S	seeds	100	Russia
Cornmint	S	aerial parts	3000	China, Brazil
Cumin	A, S	seeds	10	India
Dill	S	aerial parts	140	USA, Hungary, Bulgaria

Continued

Table 3.2 *Continued*

Oil	*Types of process used*[a]	*Plant part extracted*	*Approximate annual production* (tonnes)	*Typical country of origin*
Elemi	C, S	exudate	10	Philippines
Eucalyptus[d]	S	leaves	See individual species	
E. citriodora			800	Brazil, S. Africa, India
E. dives			50	Australia
E. globulus			1600	Spain, Portugal
E. staigeriana			50	Australia, Brazil, S. Africa
Fennel	S	seeds	80	Spain
Fir needle	S	leaves	55	Canada, USA, Russia
Galbanum	C, S	exudate	10	Iran, Lebanon, Turkey
Geranium	A, S	leaves/stems	150	Reunion, Egypt
Ginger	S	root	55	China, Jamaica
Grapefruit	E	fruit	250	Israel, Brazil, USA
Guaiacwood	S	wood	60	Paraguay
Ho	S	leaf & wood	30	China
Jasmine	A, C	flower	12	Egypt, Morocco
Juniper	S	fruit	12	Yugoslavia, Italy
Labdanum[e]	A, U, S	exudate	20	Spain
Lavender[f]	S	aerial parts	1000	France, Spain, Tasmania
Lemon	E	fruit	2500	USA, Italy, Argentina, Brazil
Lime	E	fruit	1200	Mexico, Haiti
Litsea cubeba	S	fruit	900	China
Mandarin	E	fruit	120	Italy, China
Marjoram	S	leaves and flowers	30	Morocco
Neroli[g]	S	flowers	3	Tunisia
Nutmeg	S	fruit	200	Indonesia, Sri Lanka
Oakmoss	A, C	aerial	100	Yugoslavia, Italy, France
Olibanum	C	exudate	10	Ethiopia, Yemen
Orange[g]	E	fruit	15000	USA, Brazil, Israel, Italy
Origanum	S	aerial parts	10	Spain, France
Orris	C, S	rhizome	5	Italy, France, Morocco
Palmarosa	S	leaves	55	India, Brazil
Patchouli	S	leaf	800	Indonesia
Pennyroyal	S	aerial parts	10	Morocco, Spain
Peppermint	S	aerial parts	2200	USA

Continued

Table 3.2 *Continued*

Oil	*Types of process used*[a]	*Plant part extracted*	*Approximate annual production* (tonnes)	*Typical country of origin*
Petitgrain[g]	S	leaves	280	Paraguay
Peru balsam	C, S	exudate	45	San Salvador, Brazil
Pimento	S	fruit	50	Jamaica
Pine oil	S	wood	1000	USA, Mexico, Finland, Russia
Rosemary	S	aerial parts	250	Spain, Morocco, Tunisia
Rose	C, S	flower	20	Bulgaria, Turkey, Morocco
Rosewood	S	wood	250	Brazil, Peru, Mexico
Sage	S	aerial parts	45	Yugoslavia, Spain, Greece
Sandalwood	S	wood	250	Indonesia, India
Sassafras	S	roots	750	Brazil
Spearmint	S	aerial parts	1400	USA, China, Brazil
Styrax	C, S	exudate	25	Turkey, Honduras
Tangerine	E	fruit	300	Brazil
Tarragon	S	aerial parts	10	Italy, Morocco
Thyme	S	aerial parts	25	Spain
Ti tree	S	leaves	10	Australia
Vanilla	C, T	fruit	2500 (dried beans)	Reunion, Madagascar
Vetiver	S	root	260	Reunion, Haiti, Indonesia
Ylang-ylang	A, C, S	flower	90	Comores, Madagascar

[a] A = absolutes; C = concretes and resinoids; D = dry distilled oil; E = expressed oil; S = steam-distilled oil; T = tincture; U = untreated (however, these products are often boiled to free them from plant material). [b] The wood is juniper. [c] There are two main types of cedarwood sources for perfumery. One comprises plants of the *Juniperus* family and the other of the *Cedrus* family. The former are known as English, Texan or Chinese cedarwood and are produced principally in China and the USA. The leaf oils are extracted from this family. Chemically, the major components of these oils are based on the cedrane skeleton. The products obtained from members of the *Cedrus* family are known as Atlas or Himalayan cedarwoods. They are produced in North Africa and the Himalayas and the chemical structures of their major components are based on the bisabolane skeleton. [d] The different *Eucalyptus* species of importance to the perfumery industry contain different terpenes as their major components. The terpenes of each are related to the following major components: *E. citriodora*, citronellal; *E. dives*, piperitone, *E. globulus*, cineole; *E. staigeriana*, citral. [e] The oil and absolute are known as cistus. [f] There are three species of lavender which give oils of differing quality. They are known as lavender, lavandin and spike. [g] It is interesting to note that, in the case of orange, three different oils are produced from the same species.

varies. The percentage of the individual components present in the lavender oils depends on, for example, the area where the plant was grown, the rainfall that season, the harvesting method and so on. The analytical chemist responsible for QC of lavender oil therefore does not expect to see major components present at fixed levels, but rather within an acceptable range. Linalyl acetate, for example, is normally present in lavender oil at between 30 and 60%. Synthetic linalyl acetate is available for a fraction of the price of lavender oil, and so a supplier might be tempted to add some synthetic material to the oil and charge the full price for the mixture. This fraud is also relatively easily detected. Synthetic linalyl acetate is made, as is discussed later, from dehydrolinalool. The dehydrolinalool is hydrogenated over a Lindlar catalyst to give linalool. This hydrogenation, in theory, stops at linalool. However, a small amount of the substrate is fully saturated to dihydrolinalool, which does not occur in nature. Thus, the analyst examines the gas chromotography (GC) trace of lavender oil for dihydrolinalyl acetate. If it is present, then adulteration is suspected. The tell-tale component can be detected at extremely low concentrations, using GC–mass spectrometry (GC–MS) if necessary, since that technique is even more sensitive than GC alone.

Vanilla is a very expensive natural product, costing £5000 or more *per* kilogramme. The most important components of the vanilla bean, as far as flavour is concerned, is vanillin. Synthetic vanillin costs only a few pounds per kilogramme. Food labelling laws are very strict and the penalties for declaring a flavour to be natural when it is not, are very high. Not only can companies be fined for false declaration but also their directors are liable to imprisonment in certain countries, in particular in the USA. It is therefore very important that a company buying vanilla is able to verify for itself that the goods for sale are of natural origin.

One simple test is to measure the level of radioactivity from the sample. Synthetic vanillin is not radioactive. However, natural vanilla, like all natural products, is. This is, of course, because atmospheric carbon dioxide contains some radioactive ^{14}C formed by exposure to cosmic radiation in the upper atmosphere. Plants then incorporate this into their photosynthetic pathway and produce metabolites which exhibit a low level of radioactivity. Synthetic vanillin is prepared from coal tar, which is not radioactive since the ^{14}C has long-since decayed. However, unscrupulous dealers know this and can synthesize radio-labelled or 'hot' vanillin and dose it into synthetic material so that the level of radioactivity matches that of a natural sample. Another method of checking for naturalness must, therefore, be found. When plant enzymes synthesize molecules, they, like all catalysts, are susceptible to isotope effects. The vanilla plant is no exception and examination of the distribution of hydrogen and carbon isotopes in the vanillin molecule reveals that the heavier deuterium and ^{13}C isotopes accumulate at certain specific sites. A suitable nuclear magnetic resonance (NMR) spectrometer can determine the isotopic distribution in a sample and the cost of using ^{2}H, ^{13}C and ^{14}C-labelled synthetic materials to replicate the NMR spectra and radioactivity of natural vanillin in a synthetic sample are not financially attractive. Furthermore, the ^{2}H and ^{13}C labelling patterns in the vanilla bean are different from those of other natural shikimate sources and so the NMR technique can also distinguish between vanillin from vanilla and vanillin produced by degradation of lignin. (Lignin is the structural component of wood and is therefore very inexpensive.)

As each opportunity for adulteration is blocked by analysts, the crooks seek new methods and so the QC analyst must think proactively to keep ahead.

FROM NATURAL TO SYNTHETIC

Until the middle of the nineteenth century, perfumes were largely for personal application and, furthermore, that use was restricted mostly to the wealthiest strata of society because of the cost of producing the natural materials required as ingredients. The development of organic chemistry in the nineteenth century began to make synthetic chemicals available and their use in fragrances began to grow. The incorporation of synthetics into perfumery received a huge fillip in 1921 when Coco Chanel launched her famous perfume, No 5. *Chanel N°5* owes its unique character to the inclusion of synthetic aliphatic aldehydes, 2-methylundecanal in particular. The success of this fragrance inspired other fragrance houses to experiment with synthetic materials and the modern age of perfumery was born. The synthetic materials were cheaper to produce than natural materials, thus making perfume accessible to all. Furthermore, more robust molecules could be produced which would survive in acidic, basic and even oxidizing media. Thus, it became possible to put perfume into household products in which natural oils could not be used because of degradation of their components and resultant changes in odour and colour.

The use of essential oils is also restricted by their chemical stability. Many of the components of natural oils do not survive in products such as bleaches, laundry powders and even soaps. For example, the major component in jasmine oil is benzyl acetate, which is hydrolysed in all of these products owing to their high pH (13–14, 10–11 and 9–10, respectively), and it is also susceptible to the oxidants present in the first two. The indole present in jasmine causes soap to discolour. The discovery and application of synthetic fragrance materials towards the end of the nineteenth century and throughout the twentieth was therefore a momentous event in the history of the industry. Nowadays, fragrances can be used in all the consumer goods produced for personal and household care and they can be afforded by everyone. More detail of the use and performance of fragrances in products is given in later chapters; the reason for mentioning this at this point is to highlight the importance of economic considerations.

Initially, the synthetic perfumery materials were introduced through serendipitous use of the products discovered through advances in chemical technology. For example, the nitromusks were discovered by Baur while he was working on explosives related to TNT. As techniques for isolation, characterization and synthesis of organic chemicals improved, the search for new fragrance materials became more structured. In this, the fragrance industry follows the same path as the pharmaceutical industry. The first step is to identify the materials which nature uses. Thus, the chemical components of natural oils were separated by distillation and/or chromatography and their structures determined by chemical analysis and/or spectroscopy. (Details of the application of these techniques are given in Chapter 12.) Having identified the molecular structure of an odorant, the next task is to synthesize a sample that is identical to it. Synthesis serves as the final proof of the correct determination of the structure, but it also makes it possible to produce the material without relying on the natural source. Synthetic compounds whose structures are the same as those of the natural material are referred to as 'nature identical'. This classification of materials is important in legislative terms; it is easier to obtain clearance for a nature-identical material than for one which has no natural counterpart. However, the natural materials may contain structural features which make them difficult to synthesize or susceptible to degradation in the products to which perfumes are added. The next step is therefore to synthesize materials that are close in structure but not identical to the natural one. The effect of changes of structure on the odour, and other properties, of the materials can then be studied and further analogues synthesized to produce an optimum balance of odour, performance and cost. A more detailed account of this process is dealt with in Chapter 15. For the moment, the example of the chemistry of jasmine compounds serves to illustrate the overall path from natural to synthetic material.

The components of an essential oil may be classified into three groups. Some components add little or even nothing to the odour of the oil, but may serve another purpose. For instance, they could be fixatives. The components in the second group add odour and are important in forming the total impression of the oil but, smelt in isolation, are not associated immediately with the oil. The third group of compounds are the character impact compounds. These are the materials which give the characteristic notes to the oil and which, when smelt in isolation, are instantly associated with the oil. Figure 3.5 shows a GLC trace of jasmine oil and materials of each type can be seen in it. Isophytol and benzyl benzoate have very little intrinsic odour and serve mostly as fixatives. Benzyl acetate is the major component of jasmine oil and plays a significant part in the total odour. However, it possesses a fruity note which could be, and indeed is, found in many other oils.

The character-impact compounds of jasmine are jasmone and methyl jasmonate. These two are instantly recognizable as jasmine in character and are essential to the odour of the oil. Their structures are shown in Scheme 3.10, which shows their syntheses through a common intermediate. Jasmone was first synthesized by Crombie and Harper (1956), but the synthesis in Scheme 3.10 is that of Buchi and Egger (1971). Buchi's synthesis illustrates the main problem in the synthesis of nature-identical jasmone and methyl jasmonate; that is, the inclusion of the *cis*-double bond in the side chain. The most convenient method of introducing this feature is through Lindlar hydrogenation of an acetylenic compound. In terms of total synthesis of natural products, this is a relatively trivial step and is easy to carry out on a laboratory scale. However, several synthetic steps are be required to prepare the five-carbon unit for the side chain and two more are needed to introduce and hydrogenate it. On a manufacturing scale, this leads to high process costs, especially since two of the stages involve the handling of hazardous reagents, *viz*. acetylene and hydrogen. If the side chain of jasmone is replaced by a saturated one, the synthesis is made much easier, and so dihydrojasmone is much less expensive than jasmone. Stetter and Kuhlmann's (1975) two-step synthesis of dihydrojasmone from readily available starting materials is shown in Scheme 3.11. If the endocyclic double bond and the methyl substituent

---

Figure 3.5 *GLC trace of jasmine oil. Peak A = benzyl acetate (26.7% of volatiles by relative peak area); Peak B = jasmone (3.3% of volatiles by relative peak area); Peak C = methyl jasmonate (0.6% of volatiles by relative peak area); Peak D = benzyl benzoate (11.5% of volatiles by relative peak area); Peak E =* iso-*phytol (5.6% of volatiles by relative peak area)*

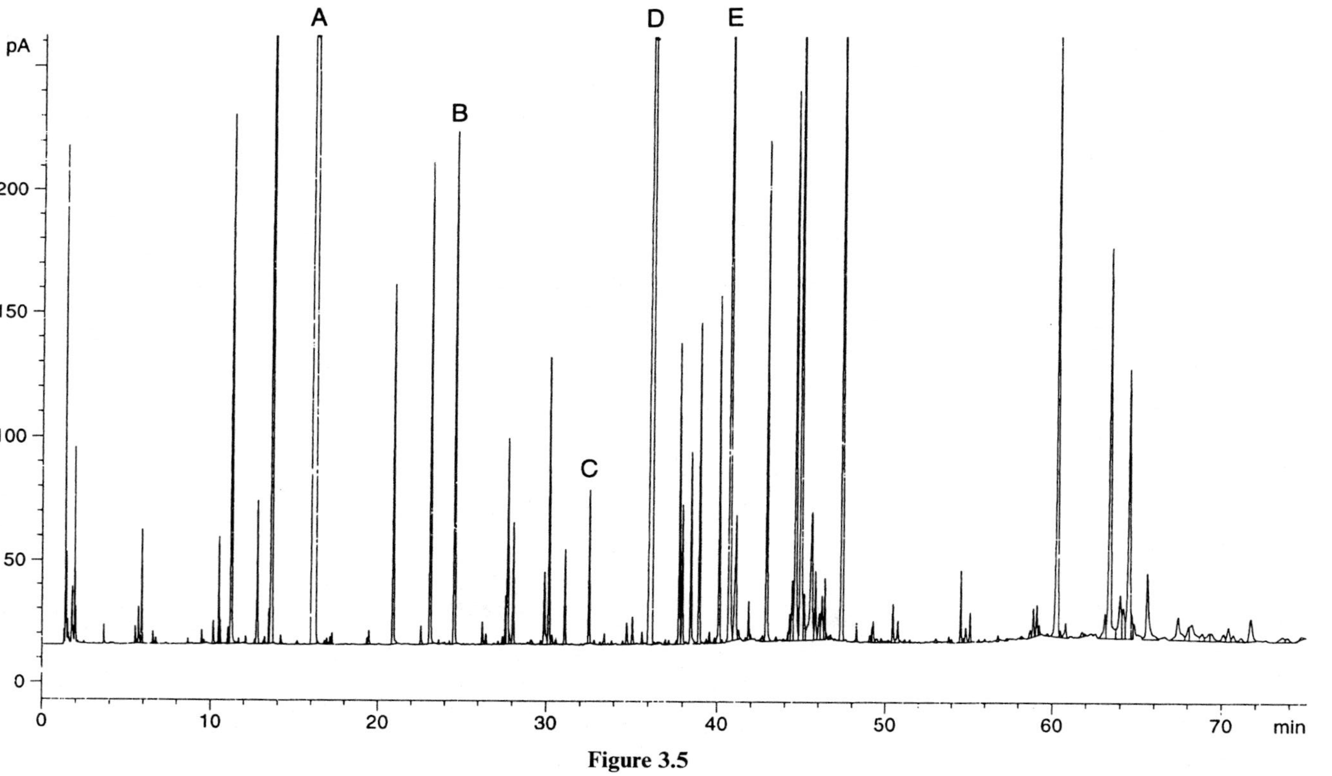

Figure 3.5

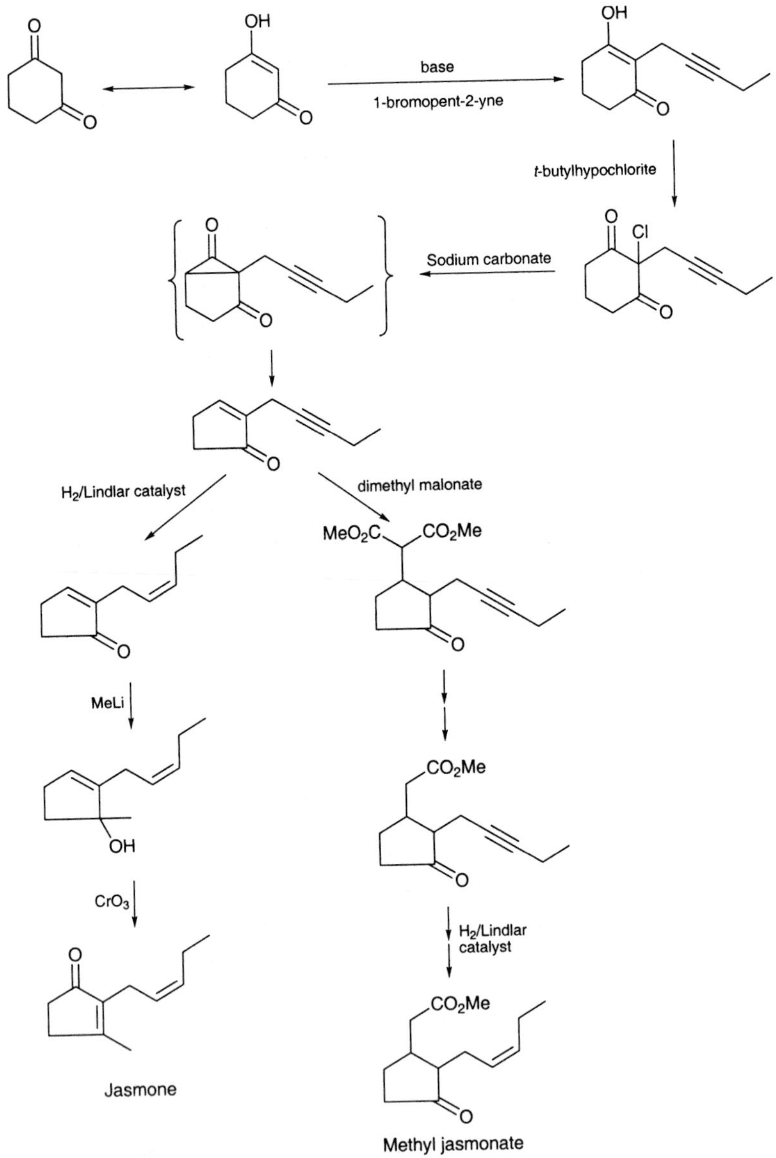

OH

base

1-bromopent-2-yne

OH

O

t-butylhypochlorite

O Cl

O

Sodium carbonate

O

O

O

H_2/Lindlar catalyst

dimethyl malonate

MeO_2C CO_2Me

O

O

MeLi

CO_2Me

OH

O

CrO_3

H_2/Lindlar catalyst

O

CO_2Me

Jasmone

O

Methyl jasmonate

Scheme 3.10

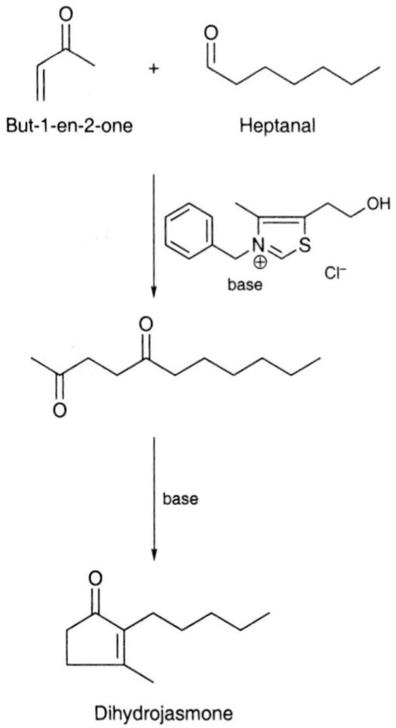

But-1-en-2-one + Heptanal

base

Cl⁻

base

Dihydrojasmone

Scheme 3.11

on the ring are also ignored, the synthesis becomes even more amenable to operation on a commercial scale (Scheme 3.12). Scheme 3.12 shows the preparation of pentylcyclopentanone, but use of different aldehydes in the initial aldol condensation gives rise to a series of homologous compounds, each with a unique blend of jasmine and fruity notes. Scheme 3.12 also shows the route used to prepare methyl dihydrojasmonate commercially. This chemistry is described in more detail in Chapter 4. Cyclopentanone is available in bulk at low cost by the pyrolysis of the calcium or barium salts of adipic acid, the precursor of Nylon 66®. This is an example of how the fragrance industry capitalizes on the availability of inexpensive feedstocks from much larger scale industries, in this case the textile industry.

Natural jasmine oils cost £3000–5000/kg, the nature-identical materials are about one tenth of that price and the price of the simpler analogues is a further order of magnitude, or even more, lower. In addition, because they lack the double bonds, the synthetic materials are more stable in products, such as laundry powder, which contain bleaching agents. All of these materials are used in fragrances, but there

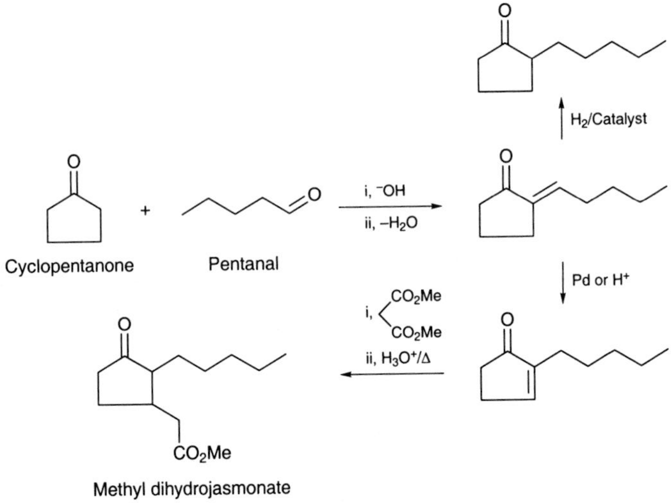

Scheme 3.12

is a correlation between price and tonnage. Obviously, the less expensive a material is, the more it is used. Jasmine absolute can only be used economically, in more than trace amounts, in the most expensive fine fragrances, whereas, 2-heptylcyclopentanone is cheap enough to allow its use in reasonably high proportions in most fragrances, including low-cost ones for use in laundry powders and household cleaners. The example of jasmine is typical of the interplay of inspiration from nature, technical possibility and economic pressure which has given rise to the variety of fragrance materials in use today and is described in Chapter 4.

REFERENCES

G. Buchi and B. Egger, *J. Org. Chem.*, 1971, **36**, 2021.
L. Crombie and S. H. Harper, *J. Chem. Soc.*, 1952, 869.
R. Croteau, Biosynthesis and catabolism of monoterpenoids, *Chem. Rev.*, 1987, **87**, 929–954.
J. D. Bu'Lock, *The Biosynthesis of Natural Products*, McGraw-Hill, 1965.
J. Mann, R. S. Davidson, J. B. Hobbs, D. V. Banthorpe and J. B. Harborne, *Natural Products: Their Chemistry and Biological Significance*, Longman, 1994.
H. Stetter and H. Kuhlmann, *Synthesis*, 1975, 379.

Chapter 4

Ingredients for the Modern Perfumery Industry

CHARLES SELL

ECONOMIC FACTORS AFFECTING PERFUME INGREDIENT PRODUCTION

The four main factors that affect the volume of use of a fragrance ingredient are its odour contribution to a fragrance, its stability and performance in the products to be perfumed, its safety in use and its cost. The first three factors are discussed in the chapters on perfumery, applications, safety and ingredient design (Chapters 7, 9, 10 and 15, respectively). The fourth factor, cost, depends on raw material availability and chemical process technology, which are discussed in this chapter.

The fragrance industry lies between the petrochemical and pharmaceutical industries in terms of scale of production and cost *per* kilogramme of product. The production scale is closer to that of the pharmaceutical industry, but the prices are closer to those of the bulk chemicals industry.

The largest-volume fragrance ingredients are produced in quantities of 5000–6000 tonnes worldwide *per annum*, but some ingredients, mostly those with extremely powerful odours, which limit their use in a fragrance, are required in only kilogramme amounts. These figures are dwarfed by products such as nylon 66 and nylon 6 (polycaprolactam), which are each produced at around 4 million tonnes annually. About 15 000 tonnes of aspirin are produced annually, and the terpenoid vitamins (A, E and K) are produced in comparable quantities to the ionone family of fragrance ingredients (with which they share a common precursor, citral).

The cost restraints on the industry are, ultimately, imposed by the consumer. In many products, such as soap and laundry powder, the fragrance may contribute significantly to the overall cost of the finished goods. If the price of one product is not acceptable, the consumer selects a competitive brand. The manufacturer of these products therefore puts considerable pressure on the fragrance supplier to develop the most effective fragrance at the lowest possible cost. The highest-volume fragrance ingredients cost only a few pounds per kilogramme. This is not far above the cost of many basic petrochemical building blocks which are, typically, in the £0.5–10/kg price range.

Thus, the fragrance chemist has to work hard and think creatively and opportunistically to provide materials at an acceptable cost, but without the advantages of scale that the bulk-chemicals industry enjoys. One method used to achieve this is to seek out materials that the bulk industries use as intermediates or produce as by-products and employ these as feedstocks for the preparation of perfume ingredients. The application of this approach is apparent in many of the sections of this chapter, in which the ingredients described are mostly based on family trees of materials produced from a common feedstock or group of feedstocks.

Experienced fragrance chemists build economic considerations into their thinking, even at the stage of designing new molecules, since they know that a material will not be successful if it cannot be produced at a competitive cost.

PERFUME INGREDIENTS DERIVED FROM TERPENES

Introduction

The terpenes form the largest group of natural odorants, so it is only to be expected that they also form the largest group of modern fragrance ingredients.

Thousands of different terpene structures occur in perfume ingredients, both natural and synthetic. The chemistry of terpenes is rich and varied and attempts to understand it have, on many occasions, contributed fundamentally to our total understanding of chemistry. One example is the work of Wagner and Meerwein, whose studies in terpene chemistry led, amongst many discoveries, to elucidation of the rearrangement that bears their names. This work made a very significant contribution to our fundamental understanding of the properties and reactions of carbocations. *Chemistry of Fragrance Substances* by Paul Teisseire (1993) serves as an excellent general introduction to the exciting field of terpene chemistry.

As far as perfume materials are concerned, the most important members of the terpene family are the oxygenated monoterpenes. The terpene hydrocarbons generally have weaker odours and are used mainly as feedstocks. The higher molecular weights of the sesquiterpenes result in their having lower vapour pressures than their monoterpene counterparts. Thus, sesquiterpenes are present at lower concentrations in the air above a perfume than are monoterpenes, with the result that to be detected they must have a greater effect on the receptors in the nose. Hence, a lower percentage of sesquiterpenes have useful odours than monoterpenes. For the same reason, very few di- or higher terpenes have odours. However, those sesqui- and higher terpenes that do have odours are very tenacious because their lower volatility means that they are lost more slowly from perfumes. Such materials form the base of perfumes and serve also to fix the more volatile components.

People in the fragrance industry tend to misuse two terms in terpene nomenclature. The word terpenes means materials whose carbon skeletons are made up from isoprene units. However, in the industry 'terpenes', is often used to mean specifically monoterpene hydrocarbons (compare the note on deterpenation of essential oils in Chapter 3). Similarly, to the chemist, geraniol is *E*-3,7-dimethylocta-2,6-dien-l-ol and nerol is *Z*-3,7-dimethylocta-2,6-dien-l-ol; whereas, in the fragrance industry, the word geraniol usually implies a mixture of the two isomers. In this chapter the terms are used in the correct chemical sense, but the reader should be aware of possible ambiguity elsewhere.

Table 4.1 shows the odour type and approximate annual consumption of some of the highest-tonnage terpene fragrance ingredients. Volumes range from these figures down to expensive specialities, which are produced in kilogramme rather than tonne quantities. The major terpene producers fall into three categories. Companies that manufacture wood and paper products produce sulfate turpentine or similar by-products rich in pinenes. Many of these companies, such as Union Camp, then produce terpenoid fragrance materials from pinenes to gain income from their by-product. One such company, Gildco, has actually split off from its parent, SCM, and become an independent aroma chemical producer. Companies (such as Hoffmann-La Roche, Bayer and BASF) that manufacture vitamins use terpene intermediates and so usually also manufacture aroma chemicals. In fact, two of these own fragrance companies; Hoffmann-La Roche own Givaudan-Roure

Table 4.1 *Some of the more important terpene fragrance materials*

Material	*Odour*	*Approximate usage* (tonnes/*annum*)
Amberlyn®/Ambrox®/Ambroxan®	Ambergris	6
Carvone	Spearmint	600
Citronellol and esters	Rose	6000
Dihydromyrcenol	Citrus, floral	2000
Geraniol–nerol and esters	Rose	6000
Hydroxycitronellal	Muguet	1000
Borneol/isoborneol and acetate	Pine	2000
Linalool	Floral, wood	4000
Linalyl acetate	Fruit, floral	3000
Menthol	Mint, coolant	5000
(Methyl)ionones	Violet	2000
α-Terpineol and acetate	Pine	3000
Acetylated cedarwood	Cedar	500

and Bayer own Haarmann & Reimer. The Japanese company Kuraray manufactures synthetic rubber from isoprene and so has diversified into terpene aroma chemical manufacture from this basic feedstock.

The remainder of this section is divided into 20 sub-sections. The first nine describe the main approaches to the production of the large-volume terpenes and the subsequent 11 describe individual groups of terpenes, sometimes classified biogenetically and otherwise, when more appropriate, by odour type. The multitude of terpene-derived fragrance ingredients makes it impossible to mention them all in the context of this book. Therefore, only a selection of the more important and interesting ones are included in this chapter.

Five Key Terpenes

Geraniol–nerol, linalool, citronellol, citronellal and citral are five of the most important terpenes as far as the perfume industry is concerned. Apart from citral, all are used as such in perfumes, and the alcohols and their esters are particularly important. All of them are key starting materials for other terpenes, as discussed later. Scheme 4.1 shows the structures of these materials and how they can be interconverted simply by isomerization, hydrogenation and oxidation. The ability to manufacture any one of these, therefore, opens up the potential to produce all of them and, hence, a wide range of other terpenes. Obviously, if one company produces geraniol–nerol initially and another linalool and both do so at the same cost *per* kilogramme, then they will not be able to compete with each other on both products. The first is restricted to geraniol–nerol and the second to linalool. Thus, the range of products that any terpene producer can market effectively depends on a fine balance of its feedstock and process costs *vis-a-vis* those of its competitors.

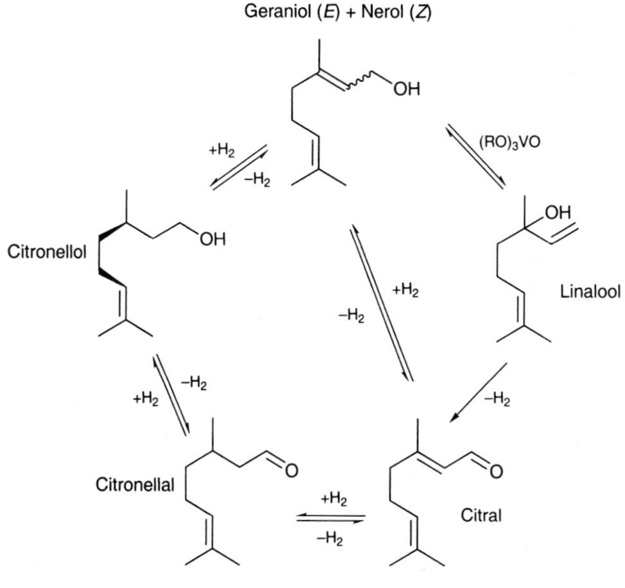

Scheme 4.1

Feedstocks

As mentioned above, routes to the major synthetic terpenes start from either turpentine or petrochemical sources.

When softwood (pine, fir, spruce) is converted into pulp in the Krafft paper process, the water-insoluble liquids which were present in it are freed and can be removed by physical separation from the process water. This material is known as crude sulfate turpentine (CST). Fractional distillation of CST gives a number of products, as shown in Table 4.2. The residue is known as tall oil and contains diterpenes.

Turpentine isolated from the Krafft process is referred to as sulfate turpentine, and that obtained by tapping living trees is known as gum turpentine. Turpentine is the monoterpene fraction of pine oil and contains mostly α- and β-pinene, present in a ratio of about 7:3 (α:β), the exact ratio depending on the species of tree involved. Pure α- and β-

Table 4.2 *Products in distillate from CST*

Product	*Percentage in CST*
Lights	1–2
α-Pinene	60–70
β-Pinene	20–25
Dipentene	3–10
Pine oil	3–7
Estragole, anethole, caryophyllenes	1–2

pinenes can be obtained by fractional distillation of turpentine. The two can be interconverted by catalytic isomerization, but this leads to an equilibrium mixture. The equilibrium can be driven in one direction by continuous removal of the lower boiling component through distillation. However, α-pinene is the lower boiling of the two and is already the more abundant. To increase the yield of β-pinene, it is necessary to fractionate, isomerize α into β, fractionate again, isomerize α into β, and so on. This is obviously a costly process in terms of time and energy. As a result of these factors concerning availability, β-pinene is about twice the price of α-pinene, which affects the economics of the processes described below.

The petrochemical building blocks used for terpene synthesis are all readily available bulk feedstocks and are therefore free from concerns regarding availability and security of supply.

Pinene Pyrolysis

One of the earliest commercial routes into this key group of terpenes involved pyrolysis of β-pinene, as shown in Scheme 4.2. When β-pinene is heated to 500 °C, the cyclobutane ring breaks *via* a *retro*-2+2 cycloaddition. This ring opening is regioselective and produces the triene, myrcene. Addition of hydrogen chloride to myrcene gives a mixture of geranyl, neryl and linalyl chlorides, which can be hydrolysed; however, the reaction with acetate anion is more efficient than that with hydroxide and so the acetate esters are usually the initial products in commercial syntheses. Fractional distillation is used to separate the various products, but is complicated by the presence of traces of chlorinated impurities and isomers formed from the opening of the cyclobutane ring in the 'wrong' direction.

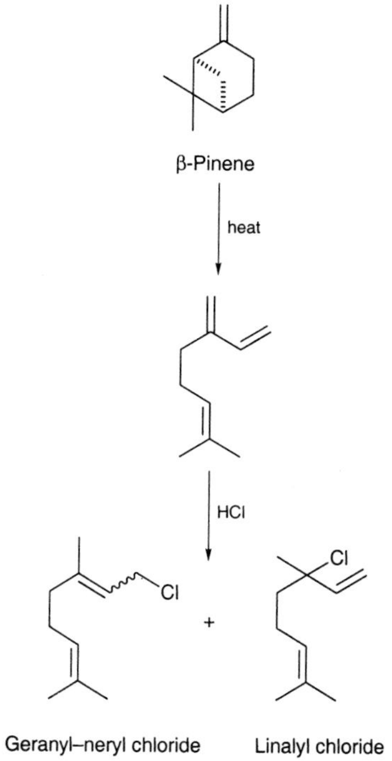

β-Pinene

heat

HCl

Cl

+

Cl

Geranyl–neryl chloride Linalyl chloride

Scheme 4.2

The two main disadvantages of this route are the cost of β-pinene and the presence of trace amounts of chlorinated materials, which must be removed from the product.

Pinane Pyrolysis

Hydrogenation of the less expensive α-pinene gives pinane, which can be oxidized by air under radical conditions to give the hydroperoxide, which is then reduced to pinan-2-ol. Pyrolysis of this alcohol gives linalool, as shown in Scheme 4.3. This process is operated by Glidco at their plant in Georgia in the USA.

The disadvantage of this process lies in a side reaction. Linalool is not stable under the pyrolysis conditions and some of it undergoes an ene reaction to give a mixture of isomeric alcohols, known as plinols. These have boiling points close to that of linalool, making separation

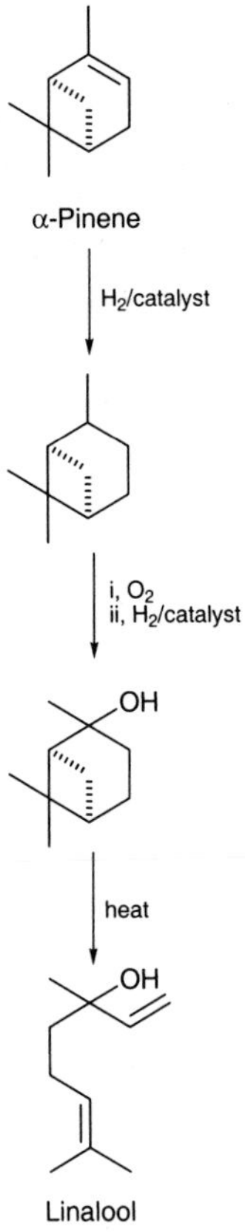

α-Pinene

H_2/catalyst

i, O_2
ii, H_2/catalyst

OH

heat

OH

Linalool

Scheme 4.3

by distillation difficult. The pyrolysis is therefore run at below total conversion to minimize plinol formation. The mechanism of the ene reaction and the structure of the plinols are shown in Scheme 4.4.

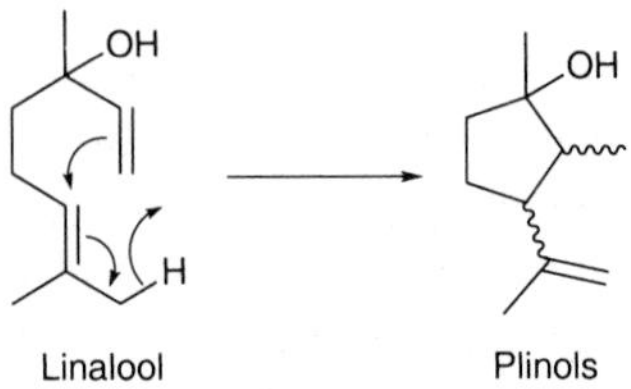

OH

H

Linalool

OH

Plinols

Scheme 4.4

The Carroll Reaction

2-Methylhept-2-en-6-one, usually referred to simply as methylheptenone, is a useful synthon for the total synthesis of terpenes. One early synthesis of this intermediate employed the Carroll reaction, the substrate for which is prepared by the addition of acetylene to acetone and subsequent partial hydrogenation to 2-methylbut-3-en-2-ol, as shown in Scheme 4.5. Addition of acetylene to methylheptenone gives dehydrolinalool, which can be hydrogenated to linalool using a Lindlar catalyst.

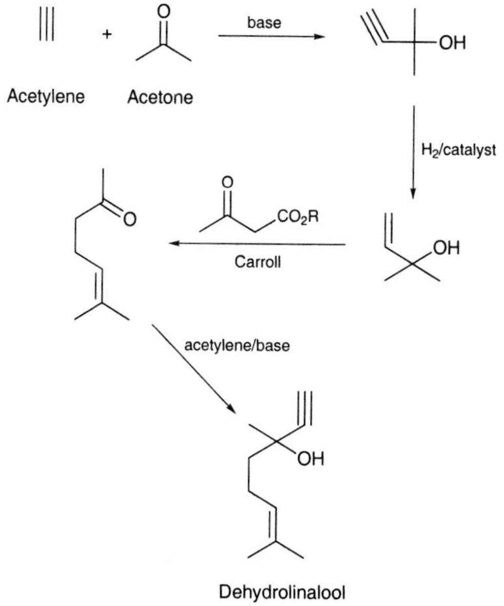

Acetylene + Acetone —base→ OH

H_2/catalyst

OH —Carroll (CO_2R)→ O

acetylene/base

OH

Dehydrolinalool

Scheme 4.5

The Carroll reaction occurs when a β-ketoester is treated with an allylic alcohol in the presence of base, or when an allyl ester of a β-ketoacid is heated. Scheme 4.6 shows the mechanism of the latter.

The disadvantage of this process is that it is not very atom efficient. The elimination of carbon dioxide means that bulk is carried through the process only to be lost towards the end, which is undesirable from both the cost and environmental impact standpoints.

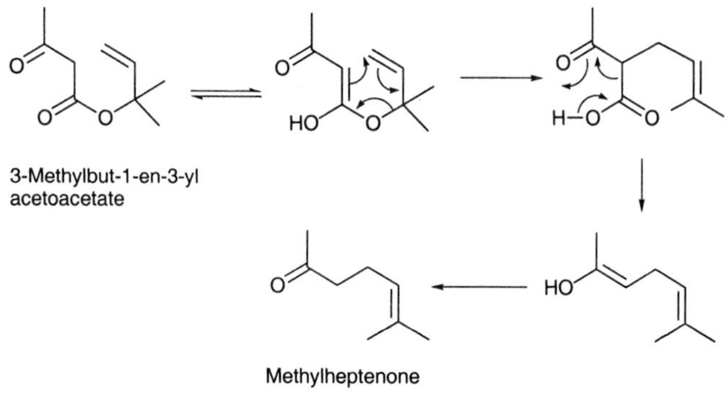

3-Methylbut-1-en-3-yl acetoacetate

Methylheptenone

Scheme 4.6

The Claisen Rearrangement

Use of the Claisen rearrangement achieves the same conversion of methylbutenol into methylheptenone as does the Carroll reaction, but without the loss of carbon dioxide. The methanol produced instead can be recovered and recycled. In this process, as shown in Scheme 4.7, methylbutenol is treated with the readily available 2-methoxypropene to give the allyl vinyl ether, which then undergoes a Claisen rearrangement to give methylheptenone.

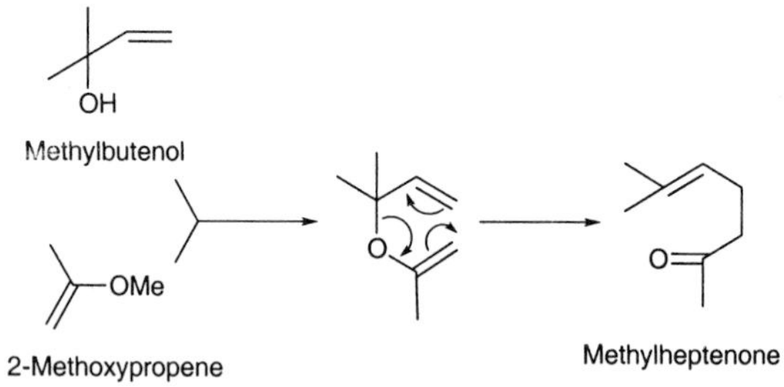

Methylbutenol

2-Methoxypropene

Methylheptenone

Scheme 4.7

Prenyl Chloride

Addition of hydrogen chloride to isoprene gives prenyl chloride, together with some of the allylic isomer 2-chloro-2-methylbut-3-ene. The presence of this tertiary chloride is not deleterious since, whilst prenyl chloride reacts with acetone *via* an S_N2 reaction, it reacts by an S_N2' mechanism to give the same product, methylheptenone (Scheme 4.8). Further elaboration to linalool, *etc*., is the same as in the processes described above. This process has been operated by Rhône-Poulenc in France and by Kuraray in Japan.

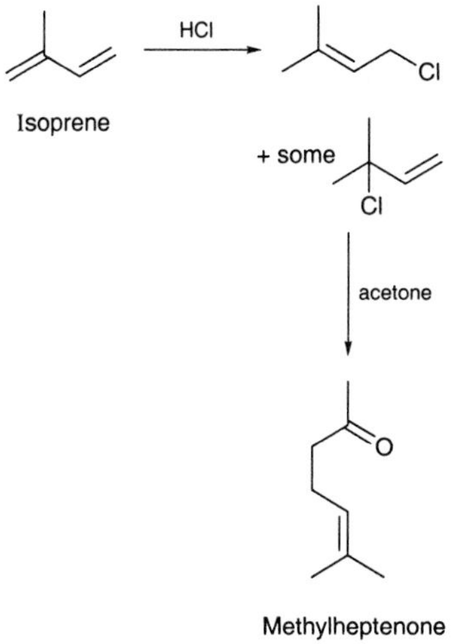

HCl

Isoprene

+ some

acetone

Methylheptenone

Scheme 4.8

The Ene Reaction

Aldol condensation of acetone with formaldehyde gives methyl vinyl ketone. This can undergo the ene reaction with isobutylene to give an isomer of methylheptenone, 2-methylhept-l-en-6-one. Isomerization to methylheptenone is easy using an acidic catalyst, and acetylene can be added to either isomer. From methylheptenone, the process leading to linalool is the same as in the examples above. If the acetylene is added to 2-methylhept-l-en-6-one, then isodehydrolinalool results. This is actually advantageous when the monoterpene unit is to be used as a precursor for ionones and vitamins, since the 1,1-disubstituted double bond is more reactive in the cyclization reaction than is the normal 1,1,2-trisubstituted bond. (The cyclization reaction is described in detail below.) The preparation of these two acetylenic ketones is shown in Scheme 4.9 and is the basis of a process commercialized by BASF.

Elegance, a Four Step Process

Another process patented by BASF is shown in Scheme 4.10. This process uses only isobutylene, formaldehyde and air as reagents; the only by-product is one molar equivalent of water per mole of product

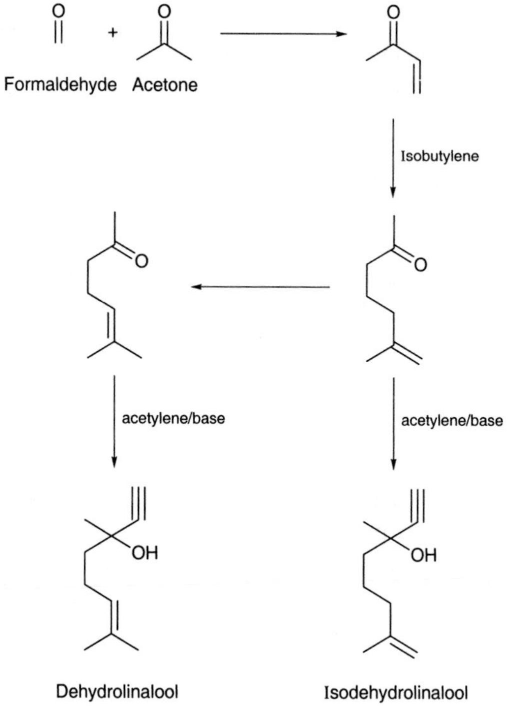
Formaldehyde
Acetone
Isobutylene
acetylene/base
acetylene/base
OH
OH
Dehydrolinalool
Isodehydrolinalool

Scheme 4.9

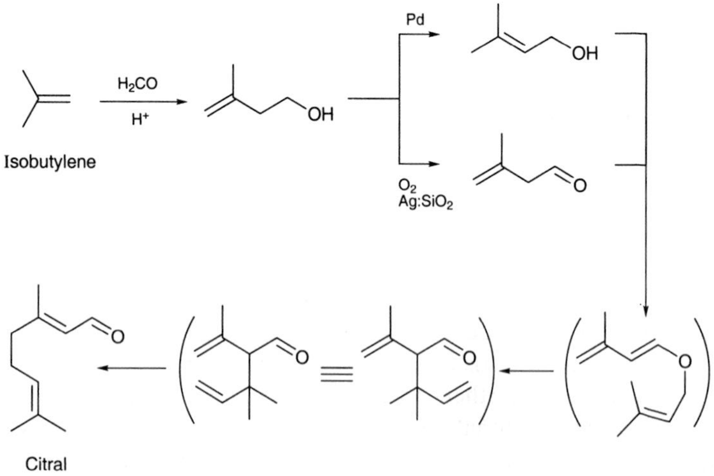
H2CO
H+
Isobutylene
OH
Pd
OH
O2
Ag:SiO2
Citral

Scheme 4.10

and it gives citral in just four steps, two of which run in parallel. Such elegance not only has intellectual appeal, but is an excellent example of how industrial chemical synthesis should be carried out, producing valuable products efficiently, at low cost and with minimal environmental impact.

The reaction between isobutylene and formaldehyde produces 2-methylbut-l-en-4-ol, isoprenol, which can be isomerized to prenol or oxidized to isoprenal. Two moles of prenol and one of isoprenal are combined to give the ketal, which then eliminates one prenol (recovered and recycled) to give the enol ether. This ether undergoes a Claisen rearrangement to give an aldehyde containing two double bonds that are perfectly placed to undergo the Cope rearrangement to give citral.

Hemiterpenes

A small number of hemiterpenes are used in perfumery, the most important of which are prenyl acetate and benzoate. Thioesters, such as those shown in Scheme 4.11, have extremely intense green odours reminiscent of galbanum, in which they occur naturally. The esters are usually prepared from prenyl chloride and the thioesters from the corresponding thiol and acid chloride.

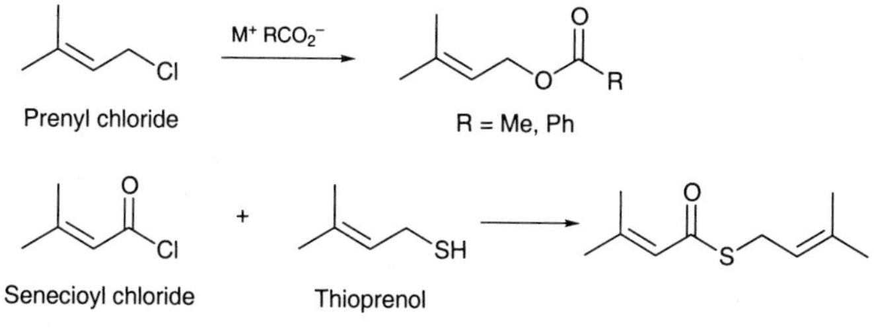

Cl
$M^+ RCO_2^-$
O
O
R
Prenyl chloride
R = Me, Ph
O
Cl
+
SH
O
S
Senecioyl chloride
Thioprenol

Scheme 4.11

Acyclic Monoterpenes

The alcohols geraniol–nerol, linalool, citronellol and their esters are the largest tonnage materials of this class. Their syntheses are described in detail above and no further explanation is necessary here. The chemical stability of these materials is limited by the unsaturation in them. To improve stability, particularly in oxidative media such as bleaches and laundry powders, various hydrogenated analogues have been developed. In some, one of the double bonds is reduced, in others both are.

A number of hydrocarbons of this family add oily, green or herbaceous notes to essential oils. Two hydrocarbons, myrcene and dihydromyrcene (also known as citronellene), deserve mention as feedstocks for other fragrance ingredients.

Pyrolysis of β-pinene gives myrcene as described above. Since it is a 1,3-diene, myrcene readily undergoes the Diels–Alder reaction with a variety of dienophiles. Addition of acrolein gives a mixture of regioisomers, the major of which is shown in Scheme 4.12. This mixture is known as Myrac Aldehyde® or Empetal®. Hydration of the double bond in the tail gives Lyral®, a widely used muguet ingredient. The reaction with 3-methylpent-3-en-2-one (from the aldol reaction of methyl ethyl ketone with acetaldehyde) is more complex in that a greater number of isomeric products are produced. Acid-catalysed cyclization gives an even more complex mixture, known as Iso E Super®. This mixture has a pleasant, woody, amber odour which is believed to arise predominantly from only a few of its components. The adduct of myrcene with methacrolein is known by the slightly misleading name of Precyclemone B®.

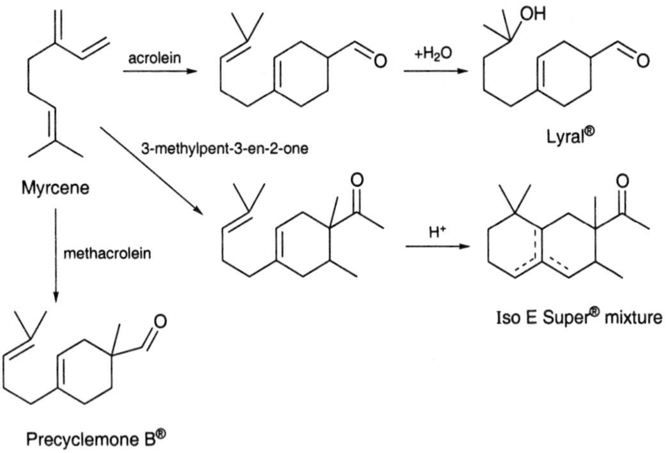

Scheme 4.12

Pyrolysis of pinane gives dihydromyrcene. Hydration of the more reactive, trisubstituted double bond gives dihydromyrcenol, as shown in Scheme 4.13. Direct hydration is difficult, so usually a two-stage process is used. The first involves the acid-catalysed addition of chloride, sulfate or acetate (from the corresponding acids), followed

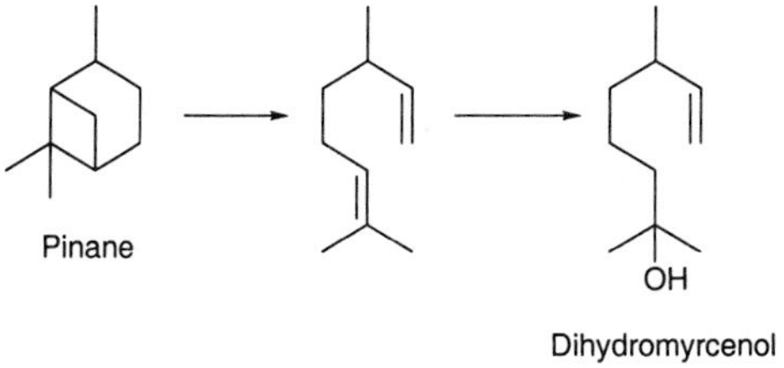

Pinane

Dihydromyrcenol

Scheme 4.13

by hydrolysis to give the alcohol. This alcohol was first introduced as a stable, fresh floral–muguet note for functional products, but its success in the after-shave *Drakkar Noir* caused a trickle-up to the fine-fragrance market, the reverse of the usual trend with new ingredients.

Hydration of the double bond of citronellal gives the compound known as hydroxycitronellal in a reaction analogous to that used to prepare Lyral®. Like Lyral®, hydroxycitronellal has a muguet odour. In acidic conditions, citronellal cyclizes to isopulegol and so the aldehydic group must be protected during the hydration stage. A typical sequence is shown in Scheme 4.14. The hydration is carried out in concentrated acid, under which conditions the oxazolidine ring is stable. Dilution of the medium allows hydrolysis of the protecting group to occur.

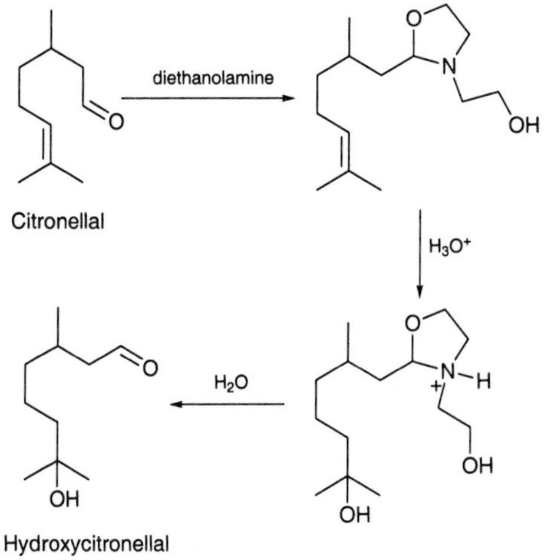

diethanolamine

Citronellal

H_3O^+

H_2O

Hydroxycitronellal

Scheme 4.14

Citral is the key odoriferous principle of lemon oil and is therefore potentially very useful in perfumery. However, it is not stable to oxidation and so cannot be used in functional products containing bleach. Since lemon is associated with cleanliness and freshness, this represents a serious challenge for household products. One of the solutions that has been found is to convert citral into the more stable nitrile, known as geranyl nitrile. Often, nitriles have odours that closely resemble the corresponding aldehyde, this being a case in point. Geranyl nitrile can be prepared from either citral or from methylheptenone, as shown in Scheme 4.15.

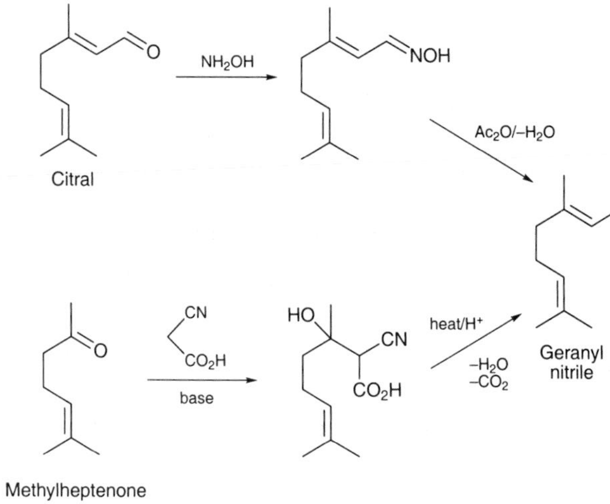

Scheme 4.15

There are several cyclic ethers derived from acyclic monoterpenes which are of importance at lower levels in fragrances. Allylic oxidation of citronellol can be used to introduce a leaving group which allows cyclization to form the pyran, rose oxide. Chlorination was one of the first oxidation techniques employed; various others, including electrochemical methods, have since been developed. An outline of the synthesis is given in Scheme 4.16. Rose oxide occurs in rose and geranium oils, to which it imparts a characteristic dry, green, rosy top-note.

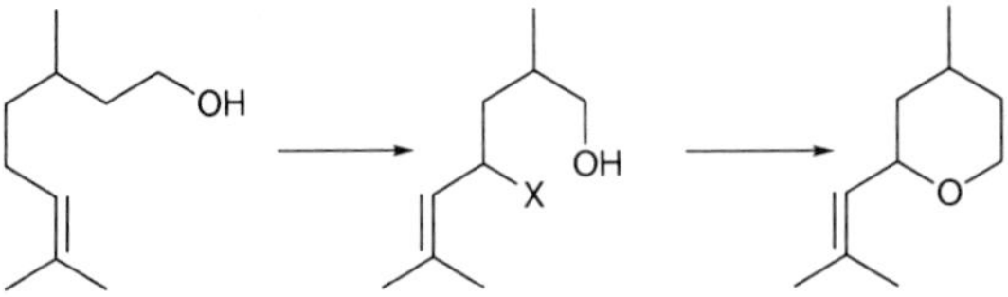

Citronellol

Rose oxide

Scheme 4.16

Structurally related to rose oxide is the hydroxypyran shown in Scheme 4.17. This material is known under the tradenames Florosa® and Florol®. It is prepared by the Prins reaction between isoprenol and isovaleraldehyde.

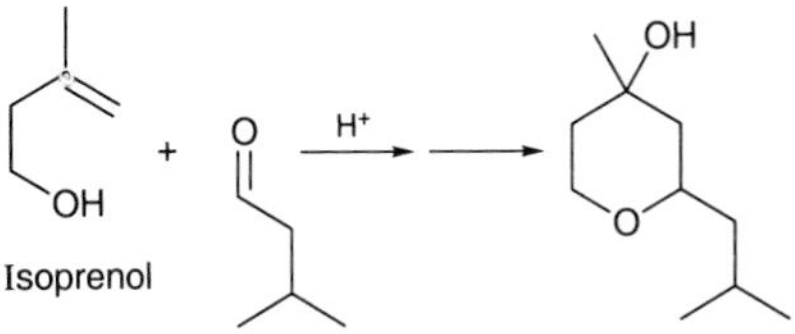

Isoprenol

Isovaleraldehyde

Florosa®/Florol®

Scheme 4.17

Oxidation of the trisubstituted double bond of linalool leads to the isomeric linalool oxides and their esters, as shown in Scheme 4.18. These materials have odours ranging from floral to woody.

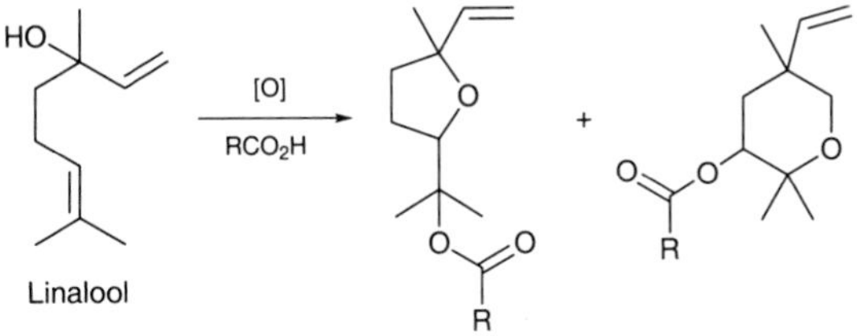

Linalool

Scheme 4.18

Cyclic Monoterpenes

The three most important cyclic monoterpenes are L-menthol, L-carvone and α-terpineol (including its esters). L-Menthol occurs in a number of mint oils and is used not only for its minty odour, but also, and more importantly, for its physiological cooling effect. Its chemistry is of such interest and significance that it warrants a section of its own (see pages 70–76).

L-Carvone is the principal odour component of spearmint oil. Both the oil and synthetic L-carvone are used as ingredients in mint flavours. The synthetic material is made from D-limonene, which is the major component of orange oil and therefore is available as a by-product of orange juice production. Quest International is the world's major producer of L-carvone and the chemistry used in the process is shown in Scheme 4.19. The chirality of the carvone is crucial to the odour, since the enantiomeric D-carvone has an odour reminiscent of dill or caraway rather than spearmint. It is therefore important that any synthesis of carvone leads to an enantiomerically pure product. It can be seen from Scheme 4.19 that both L-carvone and D-limonene owe their chirality to the carbon atom at which the isopropenyl group is attached to the cyclohexene ring. This structural feature is used in the synthesis to control the chirality of the product by controlling the regiochemistry of the reactions that take place at the opposite end of the ring.

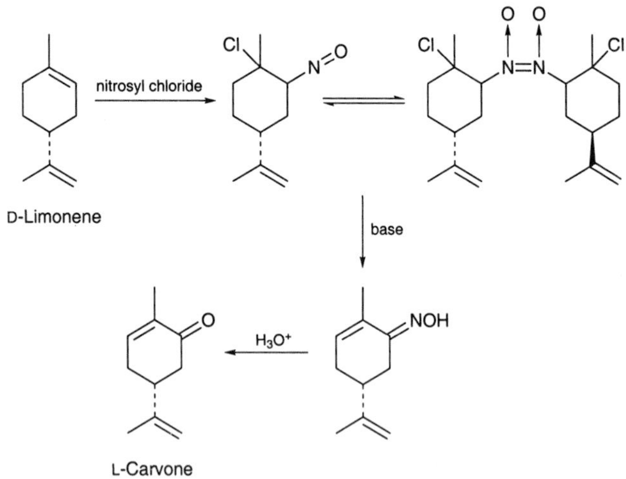

Scheme 4.19

The first step involves addition of the nitrosyl cation to D-limonene. This cation can be obtained by heterolysis of nitrosyl chloride, but in practice it is more convenient to generate it by cleavage of isopropyl nitrite using hydrochloric acid. Isopropyl nitrite is easily prepared from sodium nitrite and isopropanol. The nitrosyl cation reacts preferentially with the more electron rich endocyclic double bond of D-limonene. It also adds regioselectively to the less substituted end, thus generating the more stable tertiary carbocation. This carbocation is trapped by the chloride ions present from the hydrochloric acid. The initial adduct is a blue liquid which is in equilibrium with the dimeric white solid, as shown in Scheme 4.19. Treatment of this product mixture with a base leads to elimination of hydrogen chloride and rearrangement of the nitrosyl group to give an oxime. Thus, the reaction product is L-carvone oxime which can be hydrolysed to give L-carvone.

Treatment of turpentine with aqueous acid leads to the formation of α-terpineol. The mechanism of this reaction is shown in Scheme 4.20, in which α-pinene is used as an example. Some hydration of α-terpineol to give the diol, terpin hydrate, can also occur, the balance between the products depending on the severity of the reaction conditions. The crude mixture is known as pine oil and is the main ingredient of pine disinfectants. Terpin hydrate can be easily converted into α-terpineol, since the ring hydroxyl group is more readily eliminated than that in the side chain.

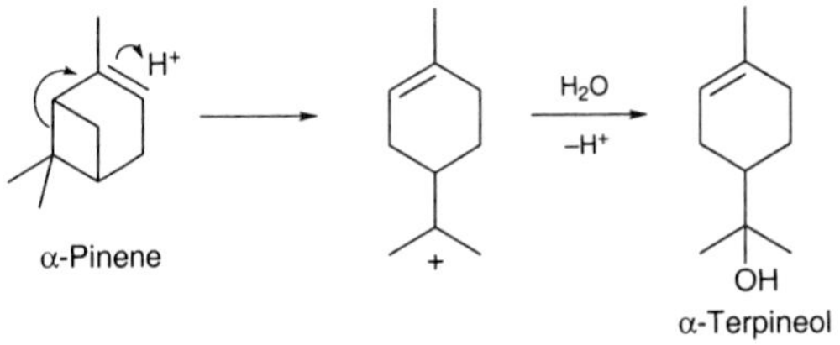

H^+

α-Pinene

H_2O

$-H^+$

OH

α-Terpineol

Scheme 4.20

Monocyclic monoterpene hydrocarbons occur in many essential oils and their by-products. They have relatively weak odours, although some add dryness and green notes to the oils containing them. This is particularly so for lime and petitgrain. D-Limonene (1) occurs in citrus oils whereas the L-isomer is found in pine. If limonene or other terpenes break down during processing to produce isoprene, then racemic limonene, dipentene, is found in the product as a result of the Diels–Alder reaction. Terpinolene (2) is the dehydration product of α-terpineol and so it is often present as an artefact. α-Phellandrene (3) occurs in eucalyptus oil. Since it is a 1,3-diene, it is an obvious precursor for Diels–Alder reactions and a number of speciality ingredients are prepared from it in this way. During processing of turpentine and other terpene sources, often a variety of acid-catalysed reactions and aerial oxidations occur. *p*-Cymene (4) is produced as a result of these processes, since it is one of the most thermodynamically stable of terpene structures. It does occur in essential oils and fragrances, but its main uses are as a thermally stable heat-transfer fluid and as a precursor for musks (see Musks, pages 91–101).

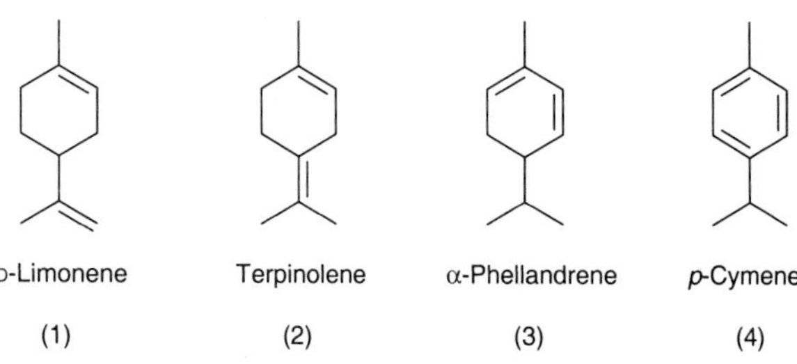

Many oxygenated monocarbocyclic monoterpenes are of use to the flavour and fragrance industry. Some are extracted from natural sources, and others are prepared from the major members of the family, usually by straightforward functional group interconversions.

Carveol (5) is one of the minor components responsible for the odour of spearmint, and is easily prepared by reduction of carvone. Isopulegol (6) is prepared from citronellal, as discussed in the section on menthol below, and is a precursor to other materials in the group. The phenols carvacrol (7) and thymol (8) are important in some herbal odour types, but the major use for thymol is as a precursor for menthol *q.v.* Piperitone (9) and pulegone (10) are strong minty odorants, the latter being the major component of pennyroyal oil. 1,8-Cineole (11) is the major component of such eucalyptus oils as *Eucalyptus globulus*. These oils are inexpensive and so there is no need to prepare cineole synthetically. Menthofuran (12) is an important minor component of mint oils and can be prepared from pulegone.

Menthol

The synthesis of menthol makes an interesting study since it neatly illustrates the balance of economic and technological factors governing the range of production methods that can be employed commercially.

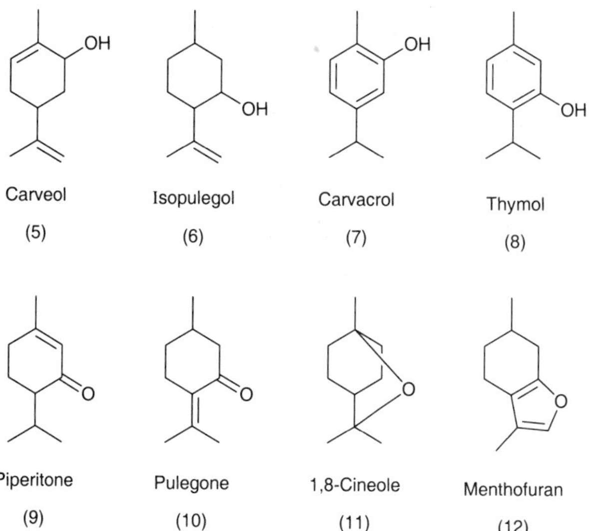

Menthol (1-methyl-4-isopropylcyclohexan-3-ol) is a monocyclic monoterpene which possesses three asymmetric carbon atoms and therefore exists in eight stereoisomeric forms (13–20). L-Menthol is the most highly desired of these since it produces a physiological cooling effect. That is, when applied to skin or mucus membranes, L-menthol creates the sensation of cooling independent of the actual temperature of the tissue concerned. It is used in toothpaste and other oral-care products, in confectionery and tobacco, and in some cosmetic products largely because of this effect. The mint taste and odour can be achieved with other materials, but the cooling effect of L-menthol is much greater than that of any of its isomers and is matched only by a few synthetic compounds (which have been found as the result of extensive research). Isomeric mixtures of menthols are less useful than pure L-menthol because the cooling effect per unit weight is lower. Therefore, any synthesis of menthol must be capable of delivering isomerically pure L-menthol to be commercially attractive.

There are three major producers of L-menthol in the world, *viz*. The People's Republic of China, Haarmann & Reimer and Takasago. Mint is grown in China and extracted to produce pure L-menthol. As a result of the vagaries of climate and competition for land from other agricultural products, the supply of natural menthol is not stable. Price and availability fluctuate and these movements have a major

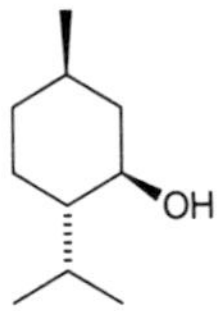

L-Menthol (13) D-Menthol (14) L-Neomenthol (15) D-Neomenthol (16)

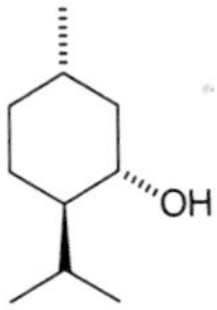

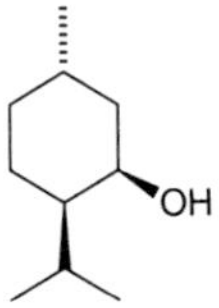

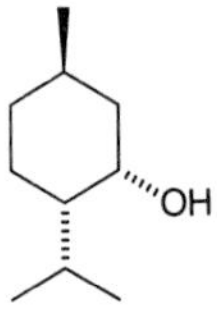

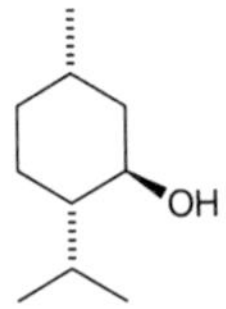

L-Isomenthol (17) D-Isomenthol (18) L-Neoisomenthol (19) D-Neoisomenthol (20)

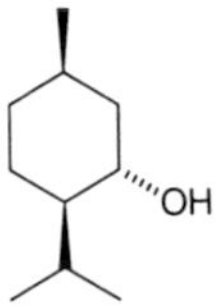

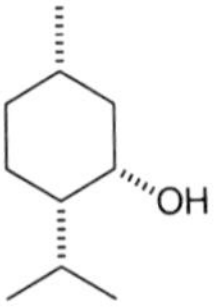

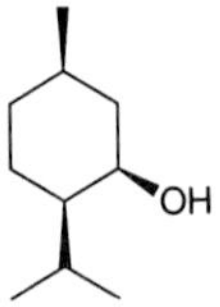

impact on the economics of the various synthetic processes for L-menthol. When natural menthol is scarce, the synthetic materials command a high price and marginal processes become economically attractive. When the natural material is in abundant supply, only the most efficient of the synthetic processes can compete. The most competitive processes are those of Haarmann & Reimer and Takasago; hence their market domination.

The company of Haarmann & Reimer was established in the nineteenth century by the two entrepreneurial German chemists whose names the company bears. The Reimer in question is the same man who gave his name to the Reimer–Tiemann reaction and, indeed, many of the company's original products were produced by the Reimer–Tiemann or similar reactions. The process that they use to produce L-menthol is shown in Scheme 4.21.

Addition of propylene to *m*-cresol produces thymol. Hydrogenation of thymol gives a mixture of menthol isomers. Treatment of any one of the eight isomers with the same copper chromite catalyst that is used for thymol hydrogenation causes racemization to the same equilibrium mixture of isomers. This fact is used to good effect in the process. The hydrogenation product is optically inactive, being composed of equal amounts of D- and L-isomers of each of the four conformational isomers. The balance between these is 62–64% menthol, 18–20% neomenthol, 10–12% isomenthol and 1–2% neoisomenthol. Since these are pairs of diastereomers, their physical properties differ. Thus,

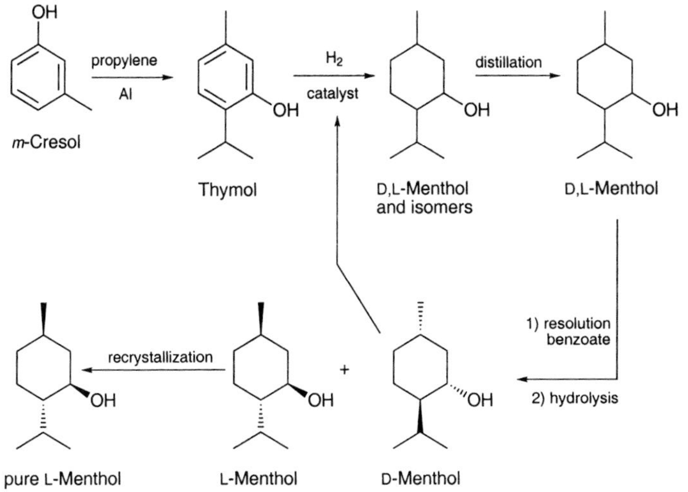

Scheme 4.21

at atmospheric pressure, D,L-menthol boils at 216.5 °C, D,L-neomenthol at 212 °C, D,L-isomenthol at 218 °C and D,L-neoisomenthol at 214.6 °C, which means that the diastereomeric pairs can be separated by distillation through a high-efficiency column and D,L-menthol obtained from the mixture. This mixture is resolved by fractional recrystallization of the benzoate ester followed by saponification. Recrystallization of the desired isomer gives pure L-menthol and all of the other seven isomers can be fed back into the hydrogenation stage with fresh thymol, where they are equilibrated as the thymol is hydrogenated. Since *m*-cresol and propylene are inexpensive feedstocks, the menthol produced by this process has a low raw material cost. However, recycling to the hydrogenation, esterification and hydrolysis and crystallization stages consumes time, labour and reactor capacity, so the low raw material cost is offset by relatively high process costs. Haarmann & Reimer are estimated to produce about 1500 tonnes of L-menthol by this process annually.

The other major producer of synthetic L-menthol is the Japanese company Takasago. They produce about 1000 tonnes per annum using elegant chemistry developed by Noyori (Scheme 4.22). Pyrolysis of β-pinene gives myrcene, to which diethylamine can be added in the presence of a catalytic amount of strong base. This produces *N*,*N*-diethylgeranylamine. Isomerization of this with the rhodium 2,2′-(diphenylphosphino)-1,1-binaphthyl (BINAP) complex produces the enamine of citronellal. The elegance of this route stems from the fact that the rhodium complex is chiral and so the proton is added to only one face of the intermediate, thus ensuring that only the enamine of D-citronellal is produced. Hydrolysis of the enamine gives D-citronellal, which can be cyclized into isopulegol by a Lewis acid-catalysed ene reaction. The chirality of the citronellal imposes itself on the transition state of the ene reaction and thus pure L-isopulegol is produced. This can be hydrogenated to L-menthol.

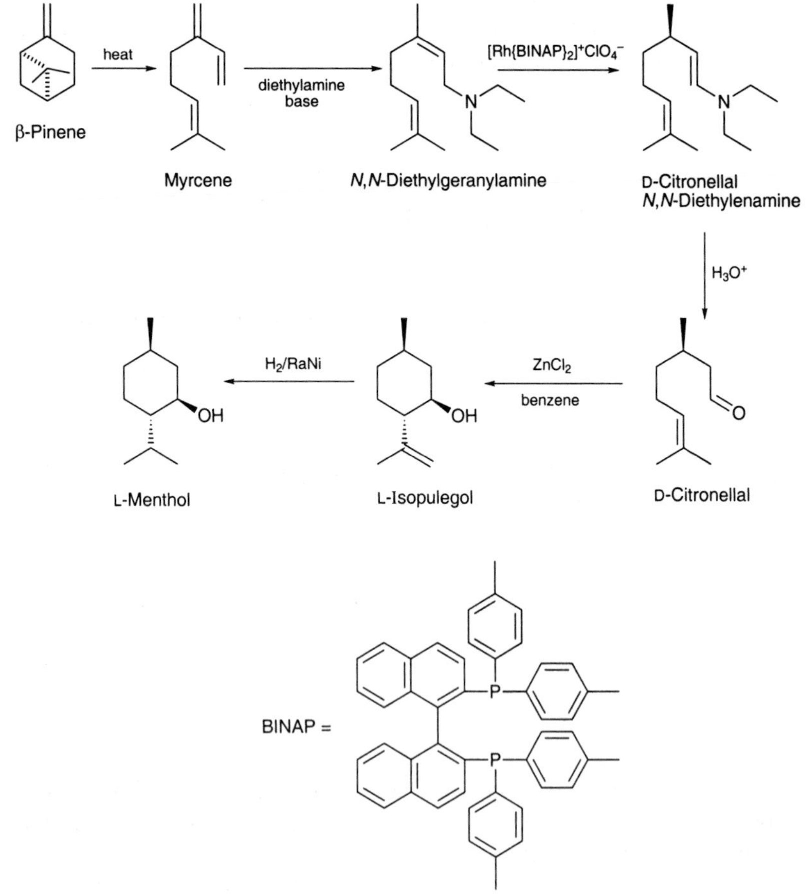

β-Pinene

heat

Myrcene

diethylamine base

N,N-Diethylgeranylamine

$[Rh\{BINAP\}_2]^+ClO_4^-$

D-Citronellal *N,N*-Diethylenamine

H_3O^+

D-Citronellal

$ZnCl_2$ benzene

L-Isopulegol

H_2/RaNi

L-Menthol

BINAP =

Scheme 4.22

The route shown in Scheme 4.23 is used by Camphor and Allied, an Indian company that has access to plentiful supplies of carene from Indian turpentine. High customs tariffs make imported menthol very expensive in India and so this process benefits from local economics. It uses the natural chirality of the turpentine-derived carene to produce D-isoterpinolene through the isomerization–pyrolysis–isomerization sequence. Hydrogenation over a poisoned catalyst gives D-3-*p*-menthene, which is epoxidized and the epoxide rearranged to give a mixture of L-menthone and D-isomenthone. Epimerization with base increases the percentage of the former in the mixture, since it is the di-equatorial (and hence more thermodynamically stable) isomer. Hydrogenation gives a mixture of isomers, the major one being the desired L-menthol, which can be separated from the D-isomers by distillation or crystallization. About 200 tonnes per annum is produced in this way.

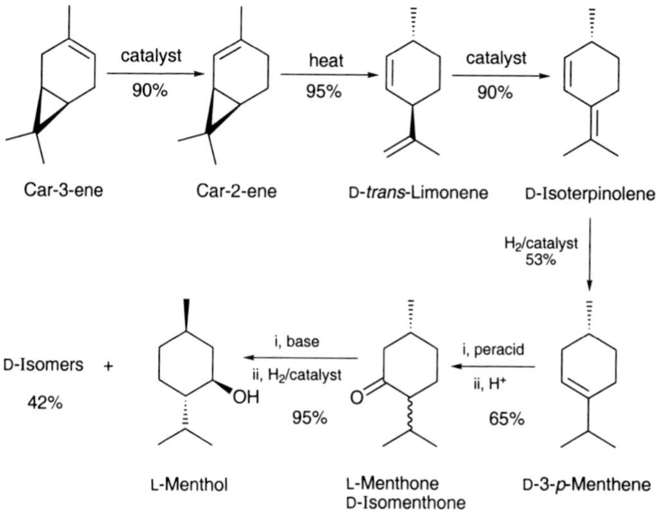

Car-3-ene —catalyst, 90%→ Car-2-ene —heat, 95%→ D-*trans*-Limonene —catalyst, 90%→ D-Isoterpinolene

D-Isoterpinolene —H_2/catalyst, 53%→ D-3-*p*-Menthene

D-3-*p*-Menthene —i, peracid; ii, H^+; 65%→ L-Menthone, D-Isomenthone

L-Menthone, D-Isomenthone —i, base; ii, H_2/catalyst; 95%→ D-Isomers 42% + L-Menthol

Scheme 4.23

Pennyroyal oil can be grown commercially in Southern Europe and North Africa. Its major constituent is D-pulegone and so another minor source of menthol relies on the chirality of this natural product to produce isomerically pure material. As shown in Scheme 4.24, hydrogenation of D-pulegone gives a mixture of D-isomenthone and L-menthone. This mixture can be separated by distillation and the

L-menthone reduced to L-menthol. The Spanish company Bordas has produced about 20 tonnes per annum by this method.

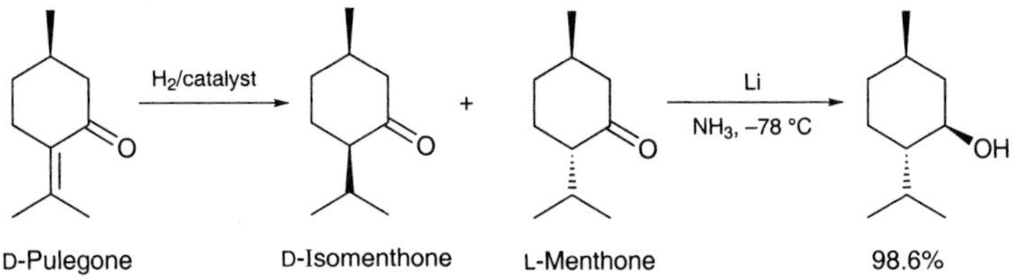

H_2/catalyst
Li
NH_3, −78 °C

D-Pulegone D-Isomenthone + L-Menthone 98.6%

Scheme 4.24

The essential oil of *Eucalyptus dives* contains L-piperitone and this provides a starting material for L-menthol using the process shown in Scheme 4.25. The L-piperitone is reduced to a mixture of piperitols, which are separated, and the major isomer, D-*trans*-piperitol, hydrogenated to give D-isomenthol containing a small amount of D-menthol. After purification, the former can be isomerized into L-menthol using aluminium isopropoxide as catalyst. About 30 tonnes per annum of L-menthol are produced by Keith Harris & Co. in Australia using this route.

This selection of menthol processes shows how the major producers are those with the most cost-effective processes, but that local economic conditions or feedstock availability can provide niche opportunities for less efficient processes.

Bicyclic Monoterpenes

Camphene can be obtained from α-pinene through acid-catalysed rearrangement. The details of the mechanism are shown in Scheme 4.26. It was study of this reaction and related conversions that led to the elucidation of the Wagner–Meerwein rearrangement. Acid-catalysed addition of acetic acid to camphene gives isobornyl acetate; the first carbocation formed in this conversion is identical to that from which camphene was formed. In the conversion of α-pinene into camphene, no good nucleophile is present and so the cation eliminates a proton. In the second reaction, the presence of acetic acid allows the cation to be trapped as the acetate. In such carbocationic rearrangements, the outcome of the reaction is usually determined by a delicate balance of steric and thermodynamic factors and by the nature of other species present in the medium. Hydrolysis of isobornyl acetate gives isoborneol, which can be oxidized to give camphor. Camphor can also be isolated from camphorwood, of which it is the major volatile constituent. The synthetic material is usually the less expensive. Camphor, isoborneol and isobornyl acetate are all used for their woody odours.

Another bicyclic monoterpene of use in perfumery is nopyl acetate, which has a sweet woody–fruity odour. Prins reaction of β-pinene with formaldehyde gives nopol which is easily esterified to the acetate (Scheme 4.27).

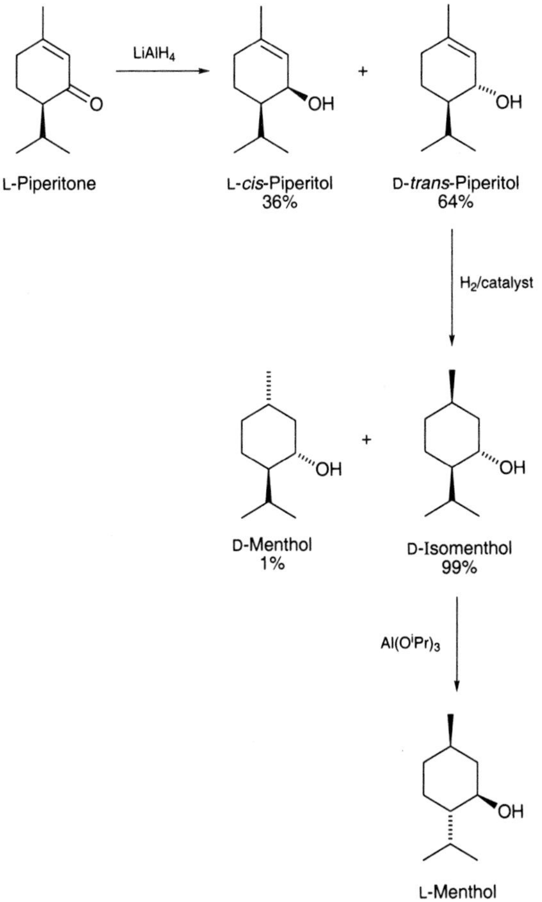

Scheme 4.25

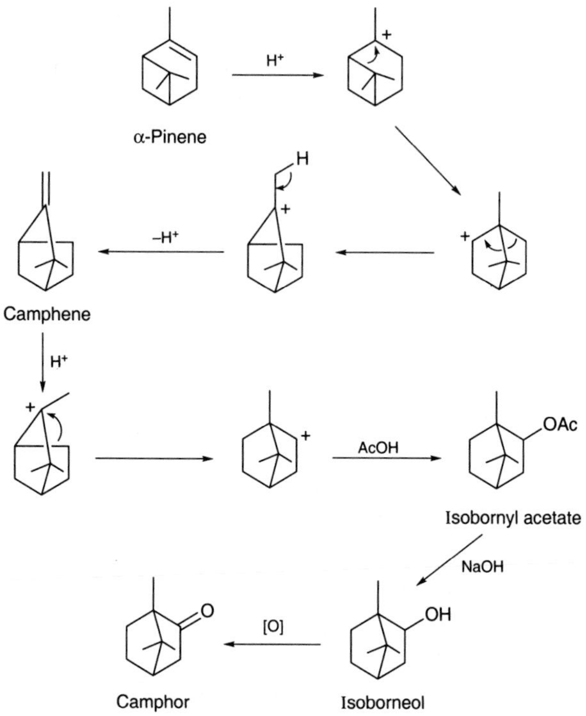
α-Pinene
H⁺
−H⁺
Camphene
H⁺
AcOH
OAc
Isobornyl acetate
NaOH
OH
[O]
O
Camphor
Isoborneol

Scheme 4.26

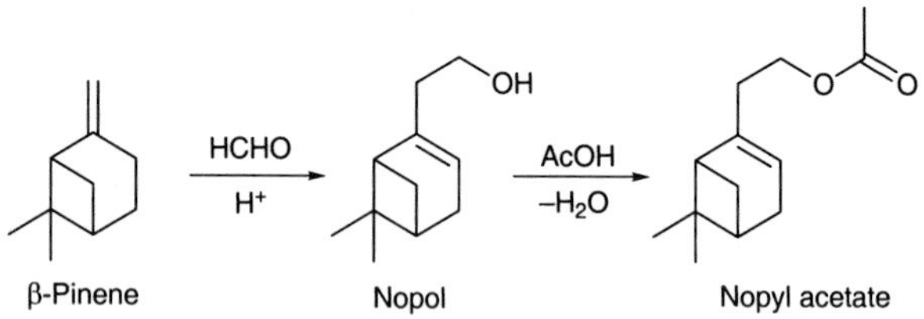
HCHO
H⁺
OH
AcOH
−H₂O
O
O
β-Pinene
Nopol
Nopyl acetate

Scheme 4.27

Sesquiterpenes

The Carroll, or an equivalent, reaction between linalool and an acetoacetic ester gives geranylacetone. Further elaboration, as shown in Scheme 4.28, produces sesquiterpenes such as nerolidol, farnesol and bisabolol which have some perfumery use, although their odours are weak. Bisabolol is used mostly because of its anti-inflammatory and

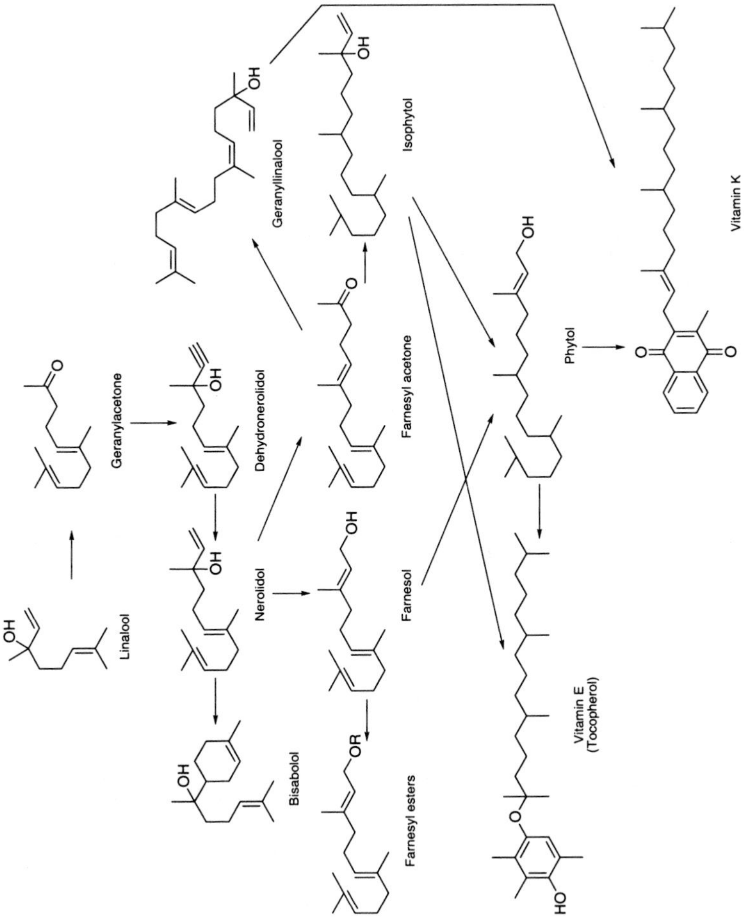

Linalool
Geranylacetone
Dehydronerolidol
Nerolidol
Bisabolol
Geranyllinalool
Farnesyl acetone
Farnesol
Farnesyl esters
Isophytol
Phytol
Vitamin K
Vitamin E (Tocopherol)

Scheme 4.28

anti-bacterial properties. The main use for these materials is as intermediates for vitamins E and K.

The sesquiterpenes important in odour terms mostly have complex cyclic structures. The problem of elimination of alcohols to produce olefins on attempted isolation is even more acute with sesquiterpenes than with monoterpenes because of their higher boiling points, which require more vigorous distillation conditions. The sesquiterpenes responsible for the odours of vetiver and patchouli oils have complex structures (21–24), which can only be reached by lengthy and hence uneconomic syntheses. As their names suggest, the vetivones occur in vetiver oil and the patchoulane derivatives in patchouli.

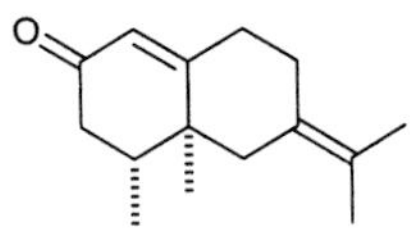

α-Vetivone

(21)

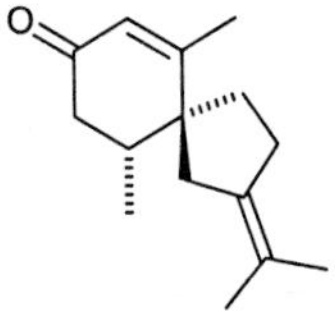

β-Vetivone

(22)

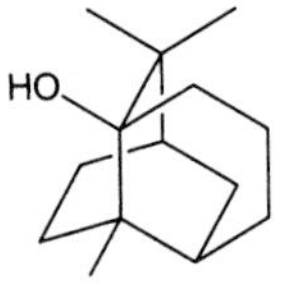

Patchouli alcohol

(23)

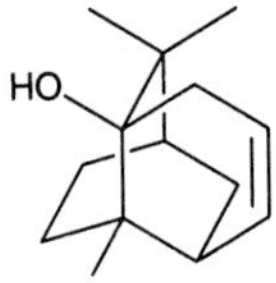

Norpatchoulenol

(24)

Caryophyllene is the main hydrocarbon component of clove oil, from which it is produced as a by-product of eugenol extraction. The endocyclic double bond of caryophyllene is highly strained and reacts readily with a variety of reagents. Usually both double bonds become involved in a *trans*-annular reaction, followed by rearrangements to give mixtures of polycyclic products. Some of these mixtures find use as woody ingredients in perfumes. Caryophyllene and two typical reaction products from it are shown in Scheme 4.29.

Longifolene is present in Indian turpentine, which is obtained from the species *Pinus longifolia*. This hydrocarbon also has a strained skeleton and treatment with acid causes rearrangement to isolongifolene. Treatment of this with a peracid under acidic catalysis gives

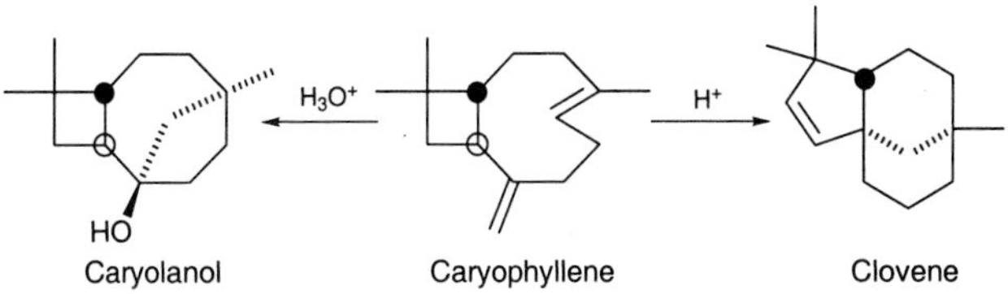

Caryolanol Caryophyllene Clovene

Scheme 4.29

isolongifolanone, which has a rich woody–amber odour. Isolongifolene can also be hydroformylated to give materials such as Amboryl Acetate®, which also has a woody–amber character. These reactions are shown in Scheme 4.30.

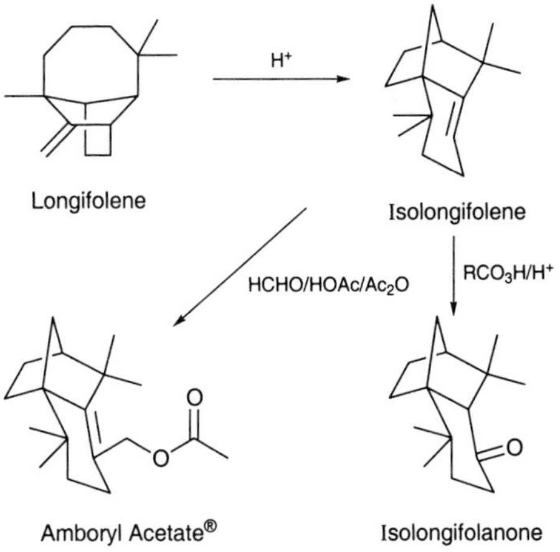

Longifolene Isolongifolene

Amboryl Acetate® Isolongifolanone

Scheme 4.30

Cedarwood

Two main families of cedarwood oils are used in perfumery. The first is extracted from trees of the family *Juniperus*. These oils are known as English, Texan or Chinese cedarwood and their components are derived from the cedrane and thujopsane groups of sesquiterpenes. Atlas and Himalayan cedarwoods are obtained from *Cedrus* species and their terpenes are mostly from the bisabolane family. The cedrane–thujopsane derivatives are much more widely used than the bisabolanes.

The major components of the *Juniperus* wood oils are cedrol, cedrene and thujopsane, and a number of perfume ingredients are made from these. The most important is acetylated cedarwood, which possesses a much stronger cedarwood odour than the natural oil; it is available under various tradenames such as Lixetone® and Vertofix®. The acetylation can be carried out using acetyl chloride in the presence of a Lewis acid or by using a Brønsted acid system such as polyphosphoric acid–acetic anhydride or sulfuric acid and acetic anhydride. The cedrol may be removed before acetylation or allowed to dehydrate to give cedrene under the reaction conditions and so the major component in the product is acetylcedrene. However, the main contributor to the odour of the complex reaction product mixture is the ketone derived from acetylation of thujopsene (Daeniker *et al.*, 1972), These reactions are shown in Scheme 4.31.

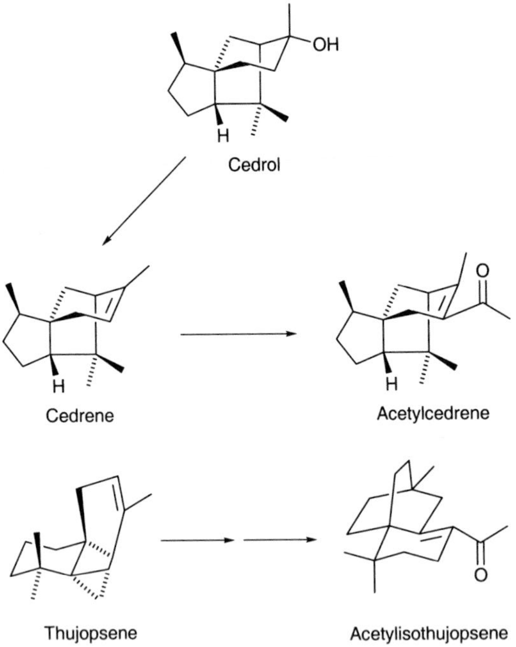

Cedrol

Cedrene

Acetylcedrene

Thujopsene

Acetylisothujopsene

Scheme 4.31

Sesquiterpene chemistry is always full of surprises. For example, an attempt to use titanic chloride in place of aluminium chloride to produce acetylcedrene resulted in a previously unknown compound (McAndrew *et al.*, 1983). In this case, as shown in Scheme 4.32, the carbocation produced by addition of the acetyl cation to cedrene, instead of eliminating a proton as normal, underwent a Wagner–

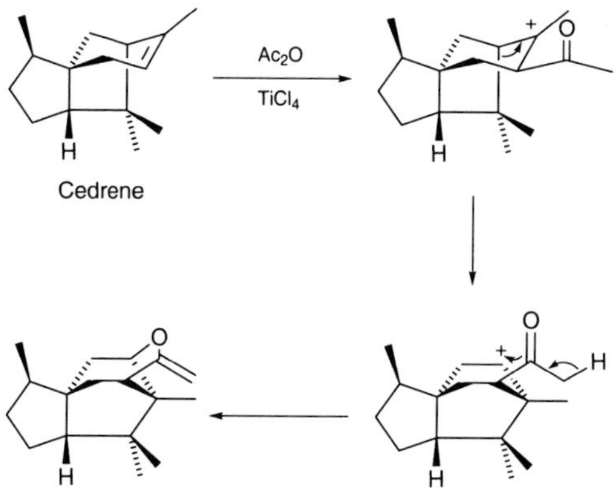

Ac2O
TiCl4
Cedrene

Scheme 4.32

Meerwein rearrangement to a second cation, which was then trapped by *trans*-annular addition of the carbonyl group.

The methyl ether of cedrol is also used as a woody–amber note. The Atlas and Himalayan cedarwoods have a sweeter and less ambery woody smell than the materials described above. This odour is due to materials such as atlantone (25) and deodarone (26).

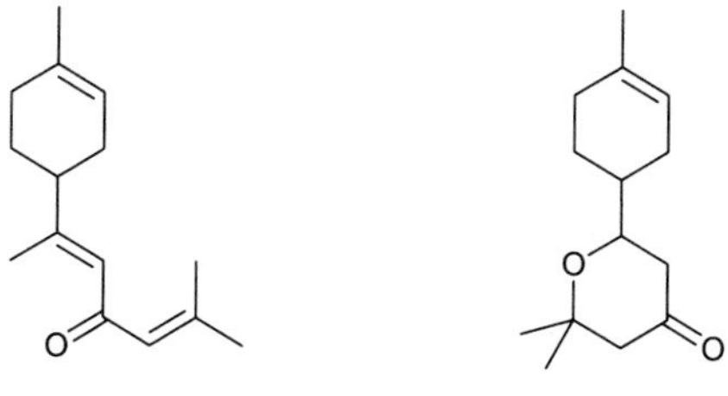

Atlantone
(25)

Deodarone
(26)

Sandalwood

Sandalwood oil is obtained by distillation of the parasitic tree *Santalum album*. The major components of the oil are the santalols (27, 28). Many syntheses of these and related chemicals have been reported but, elegant as they are, none compete economically with the oil itself. The synthetic sandalwood materials fall into two main classes, the so-called terpenophenols and the materials derived from campholenic aldehyde.

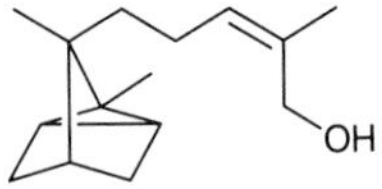

α-Santalol
(27)

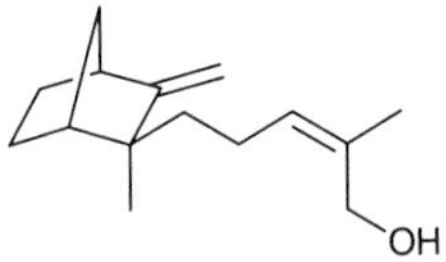

β-Santalol
(28)

Addition of camphene or borneol to guaiacol gives a mixture of products. These vary in the nature of the terpene unit (since many rearrangements are possible) and in its position of attachment to the aromatic ring. Hydrogenation under high pressure produces an even more complex mixture. Hydrogenation leads to hydrogenolysis of one of the oxygen atoms and a variety of positional and conformational isomers of the resultant substituted cyclohexanol. In all, over 100 isomers are formed, but only two are thought to contribute to the odour; those with an isobornyl group attached to the 3-position of the cyclohexanol in a *trans* configuration relative to the hydroxy group as shown in Scheme 4.33. The odour quality of the mixture depends on the exact balance of all of the components; and there are many products on the market with tradenames such as Sandela®, Sandel N®, Santalix®, Santalidol® and Indisan®.

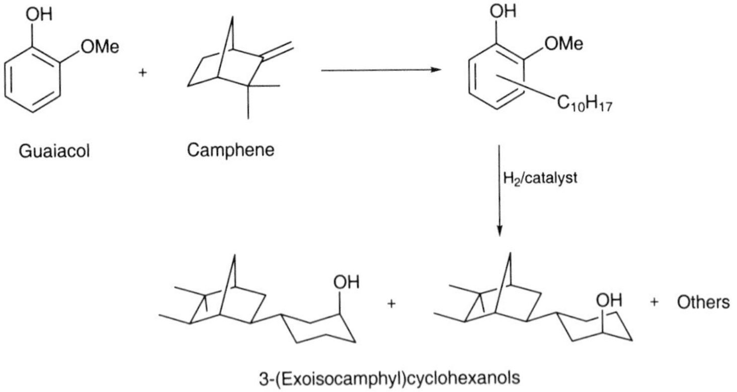

Guaiacol + Camphene → $C_{10}H_{17}$-substituted guaiacol

H_2/catalyst

3-(Exoisocamphyl)cyclohexanols + Others

Scheme 4.33

Rearrangement of α-pinene oxide, catalysed by zinc chloride or bromide, gives a product known as campholenic aldehyde (Scheme 4.34). Aldol condensation of this with a second aldehyde or a ketone gives an α,β-unsaturated carbonyl compound that can then be reduced

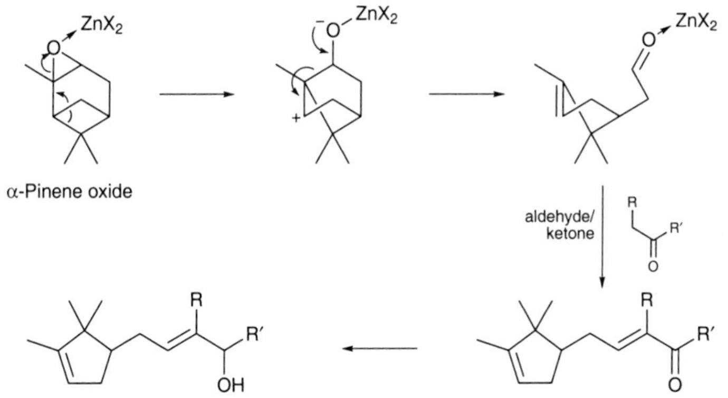
ZnX$_2$
α-Pinene oxide
ZnX$_2$
ZnX$_2$
aldehyde/
ketone
R
R′
O
R
R′
OH
R
R′
O

Scheme 4.34

to an allylic alcohol. Many materials of this type are on the market, some of which have also undergone further modification. Some tradenames, are Bangalol®/Bacdanol® (29), Brahmanol® (30), Sandalore® (31), and Polysantol® (32).

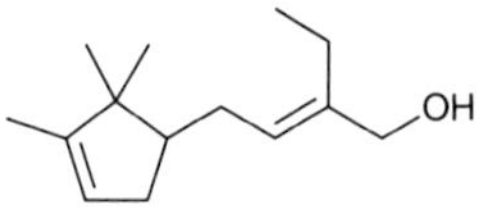
OH

Bangalol®
Bacdanol®
(29)

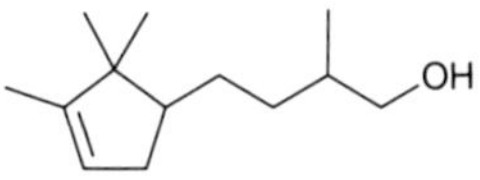
OH

Brahmanol®
(30)

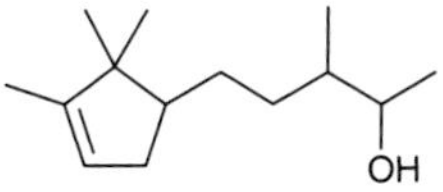
OH

Sandalore®
(31)

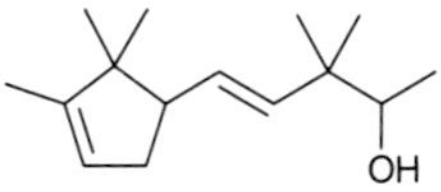
OH

Polysantol®
(32)

One unusual structure for a synthetic sandalwood is that of Osyrol®, which is produced from dihydromyrcene as shown in Scheme 4.35.

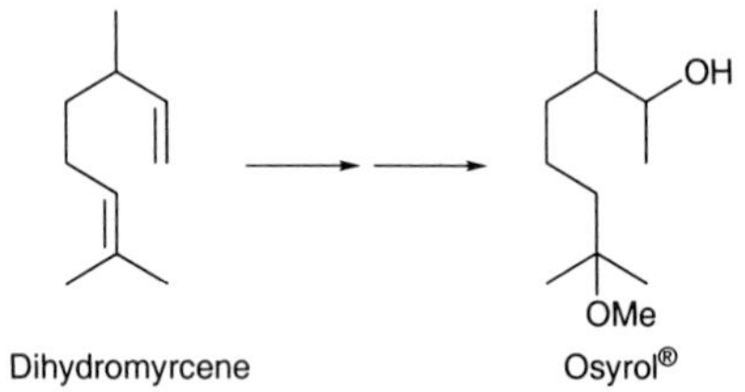

OH

OMe

Dihydromyrcene

Osyrol®

Scheme 4.35

Diterpenes

Diterpenes have, by definition, 20 carbon atoms in their structure. This means that very few are sufficiently volatile to possess an odour. One diterpene is used in perfumery because it and the derivatives concerned are odourless. That is, they are used as solvents. In view of their hydrophobicity and low volatility, these solvents also have fixative properties. Abietic acid is a major component of tall oil, the residue from distillation of turpentine. Esterification and hydrogenation produces two solvents, as shown in Scheme 4.36.

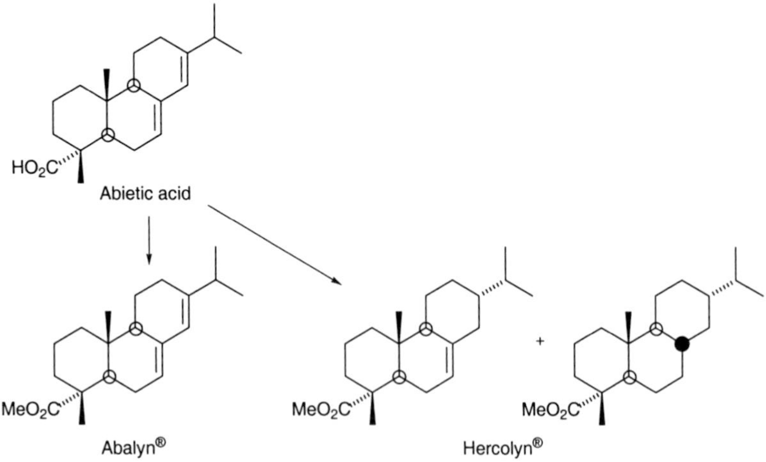

HO_2C

Abietic acid

MeO_2C

Abalyn®

MeO_2C

+

MeO_2C

Hercolyn®

Scheme 4.36

Ambergris

The last two sections deal with two very important groups of terpenoid fragrance ingredients which arise, in nature, from degradation of larger units.

The sperm whale produces, in its intestinal tract, a triterpene called ambreine, the structure of which is shown in Scheme 4.37. It is not known exactly why the whale produces ambreine; it is possibly in response to some irritation. Lumps of ambreine, which can weigh up to 100 kg, are excreted into the sea. There, in the presence of salt water, air and sunlight, the ambreine undergoes a variety of degradation reactions to produce a complex mixture of breakdown products. The mixture is known as ambergris, from the French *ambre gris* (grey amber). This name arises because ambergris is found washed up on beaches, as is amber (*ambre brun*, brown amber), the fossilized resin, which ambergris resembles to some extent. Some of the more organoleptically important degradation products are shown in Scheme 4.37. The most important of all is the perhydronaphthofuran, which possesses the characteristic animalic note of ambergris.

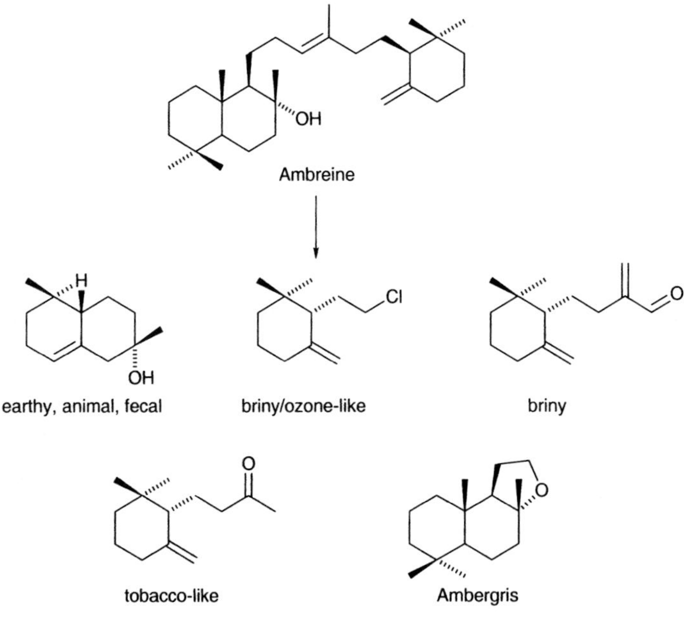

OH
Ambreine

H
OH
earthy, animal, fecal

Cl
briny/ozone-like

O
briny

O
tobacco-like

O
Ambergris

Scheme 4.37

Ambergris has always been very expensive and the decline in the whale population has exacerbated the situation. The price and availability of the natural material essentially preclude its use in fragrance and so, much work has been done on synthetic substitutes. The naphthofuran is prepared from sclareol, a diterpene found in clary sage (*Salvia sclarea*). Clary sage oil is used in perfumery and sclareol is extracted from the distillation residues. Degradation of sclareol using permanganate gives the lactone sclareolide, as shown in Scheme 4.38. Reduction of the lactone with lithium aluminium hydride, borane or some similar reagent gives a diol which can be cyclized to the furan. This is known commercially under tradenames such as Amberlyn®, Ambrox® and Ambroxan®. In addition to the nature-identical materials, there are, on the market, a number of synthetic ambergris substitutes such as Karanal®, which is described on pages 119 and 120.

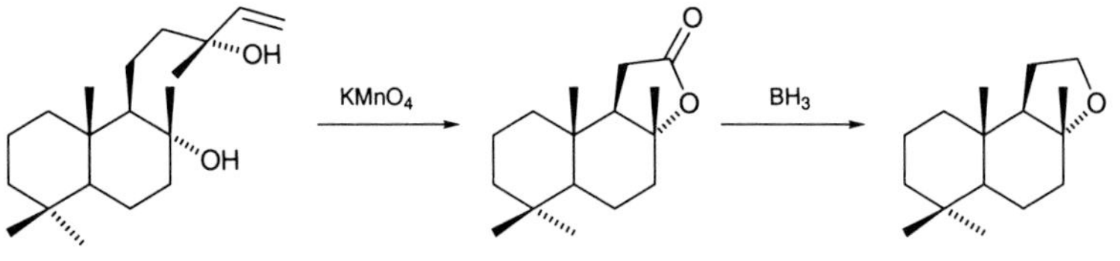

$KMnO_4$ BH_3

Sclareol Sclareolide

Scheme 4.38

Ionones and Related Compounds

The ionones and damascones are derived in nature from the degradation of carotenoids. Similarly, the related irones are formed by degradation of other higher terpenes. The ionones are synthesized from citral by aldol condensation with a ketone to form what are known as ψ-ionones, which are then cyclized using an acid catalyst, as shown in Scheme 4.39. Some specific syntheses are shown later in Scheme 4.42, along with the syntheses of vitamin A and carotene. The ionones possess odours which are reminiscent of violet, sometimes also with woody notes.

The nomenclature system used for ionones is shown in structure 33. The Greek letters α-, β- and γ- refer to the position of the double bond after ring closure. The prefixes *n*- and iso- refer to the position of the alkyl group if a higher ketone than acetone is used in the aldol condensation. For example, the ionone which is produced in greatest quantity is known as α-isomethylionone (34). (It should be noted that some older texts erroneously refer to this material as γ-ionone.) With

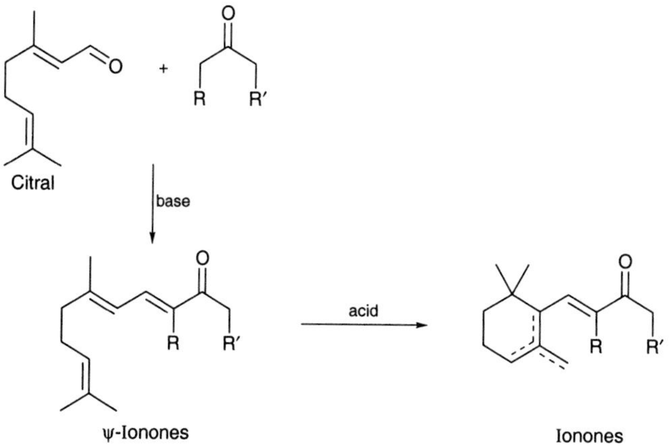

Citral + base → ψ-Ionones → acid → Ionones

Scheme 4.39

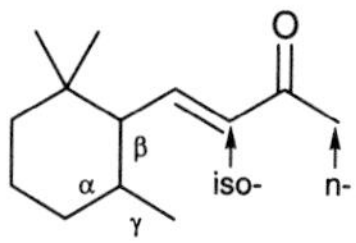

Ionone nomenclature
(33)

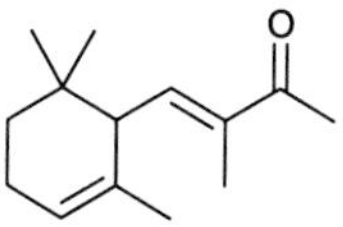

α-Isomethylionone
(34)

any given feedstock, the reaction conditions can be manipulated to produce a desired isomer predominantly. It is important to maintain strict control of process and distillation conditions to ensure reproducible odour quality. Quite a range of ionone derivatives are used in perfumery, each having its own particular odour character.

The damascones are a group of materials related to the ionones, but in which the enone unit is transposed. They are components of rose oils and have very intense fruity–floral odours. The damascenones, which contain a second double bond in the ring, are also very much sought after. One synthesis of α-damascone from methyl α-cyclogeranate is shown in Scheme 4.40. A double Grignard addition gives the tertiary alcohol. Heating of this in the presence of base, to prevent elimination

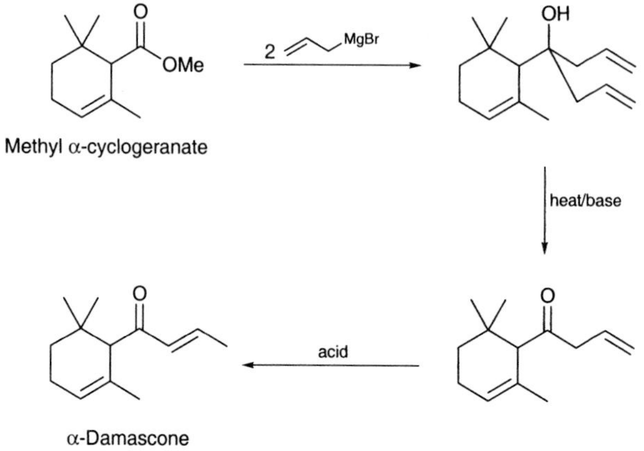

Methyl α-cyclogeranate

2 MgBr

heat/base

acid

α-Damascone

Scheme 4.40

of the alcohol, gives the ketone *via* an ene reaction. The prefixes α-, β-, and γ- have the same meaning in the damascone series as in the ionones.

The irones are higher analogues of the ionones, in that they contain an additional methyl group in the cyclohexane ring. They are components of orris and have odours reminiscent of that source. One example of a synthesis of the irones, starting from pinene, is given in Scheme 4.41.

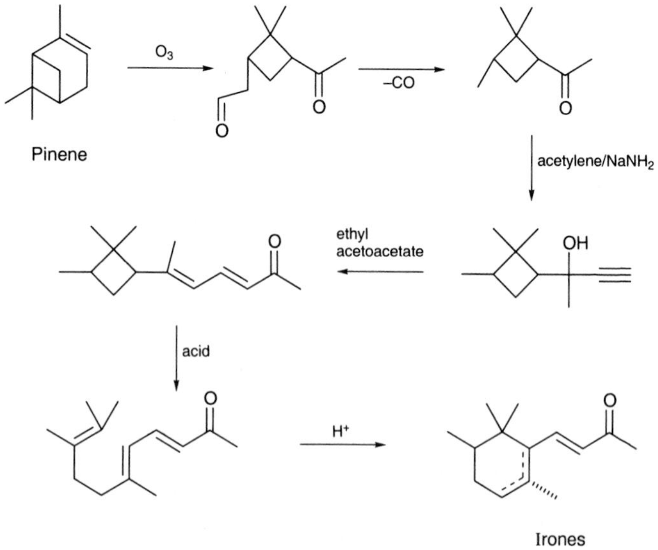

Pinene

O_3

–CO

acetylene/$NaNH_2$

ethyl acetoacetate

acid

H^+

Irones

Scheme 4.41

Decarbonylation of the ozonolysis product of α-pinene gives a methyl trimethylcyclobutyl ketone. Addition of acetylene to this, followed by the Carroll reaction, gives (trimethylcyclobutyl)heptadienone. The cyclobutane ring is opened by acid to give a methylated ψ-ionone which, on cyclization, produces irones. Once again, the Greek prefixes have the same significance as in the ionone series.

It is obvious that the syntheses of the damascones and irones are more difficult than those of the ionones, with the inevitable result that these are much more expensive products.

The economics of ionone production is affected by the fact that ionone is an intermediate for vitamin A and carotene production. However, the volumes of production of ionones are not far from those of the A vitamins and so, while some companies produce both, others can compete effectively even though they are only active in one. Scheme 4.42 shows the routes used to produce vitamin A and carotene from the various monoterpene precursors.

MUSKS

The original musk components of perfumes were extracted from animal sources. The two major ones were musk and civet, which were extracted from the anal glands of the musk deer and the civet cat, respectively. In both cases, the extracts have a strong animalic character arising from compounds such as indole and skatole (2-methylindole). The musk deer (*Moschus moschiferus*) is found in central Asia and natural musk is often called musk Tonkin, to denote the region of its origin. The civet cat (*Viverra civetta*) is found in Africa and Asia. Somehow, early perfumers recognized that under this animalic, and not entirely pleasant, note lay a rich, sweet character which serves not only as a perfume component in its own right, but also to 'fix' other notes. Fixation is a property of some perfume components, usually the higher boiling ones, that enables them to fix or hold back the more volatile notes so that they evaporate less quickly. The fixatives are therefore important in blending all of the perfume components so that the character does not change rapidly as each ingredient evaporates in turn.

Muscone (35) and civetone (36) are the most important odour components of musk and civet respectively. Ambrettolide (37) is a plant product, occurring in the seeds of the ambrette plant (*Abelmoschus moschatus* Moench, syn. *Hibiscus abelmoschus*) which is cultivated in Madagascar, the Seychelles, Colombia and Equador.

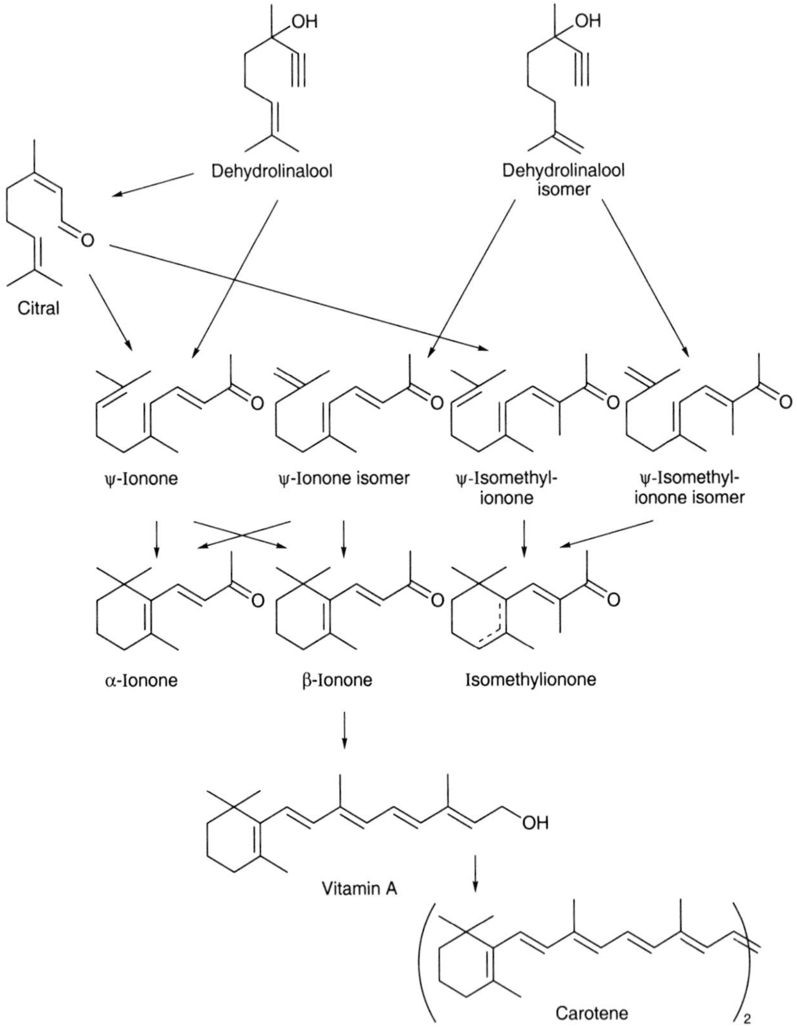

OH
OH
Dehydrolinalool
Dehydrolinalool isomer
O
Citral
O
O
O
O
ψ-Ionone
ψ-Ionone isomer
ψ-Isomethyl-ionone
ψ-Isomethyl-ionone isomer
O
O
O
α-Ionone
β-Ionone
Isomethylionone
OH
Vitamin A
Carotene
2

Scheme 4.42

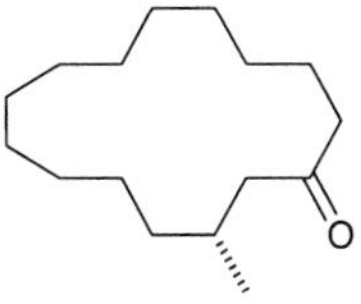

Muscone

(35)

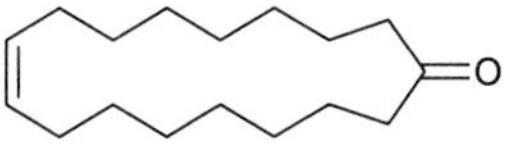

Civetone

(36)

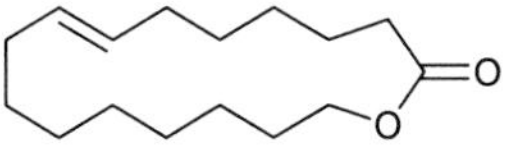

Ambrettolide

(37)

Nitromusks

The natural musks were always very expensive and their macrocyclic structures presented synthetic challenges which were not conquered, even on laboratory scale, until the pioneering work of Ruzicka in 1926. It was therefore of major importance to the fragrance industry when, in 1888, Baur discovered the nitromusks. He had actually been working on explosives and noticed that the product of *t*-butylation of trinitrotoluene (TNT) had a pleasant, sweet, musky odour. The compound was named Musk Baur® (38), although the alternative name, Musk Toluene®, eventually became more common. For a while it was also known as Tonkinol® because of the similarity of its odour to that of musk Tonkin. Baur then searched for analogues of this material and

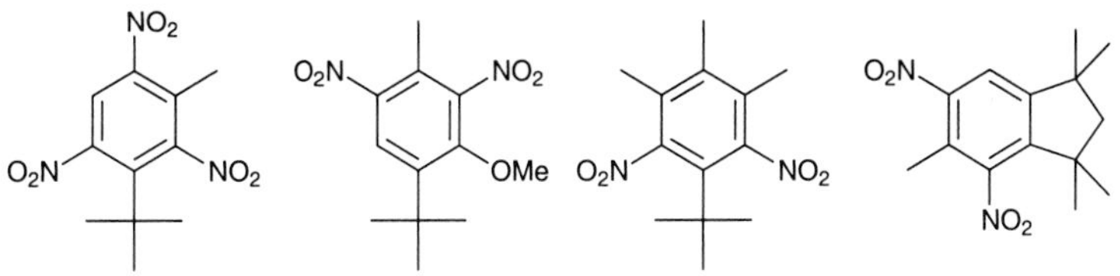

Musk Baur® (38) Musk Ambrette® (39) Musk Tibetine® (40) Moskene® (41)

discovered Musk Xylene® and Musk Ketone® (synthetic routes shown in Scheme 4.43) and Musk Ambrette® (39). Musk Ketone® is considered to have the closest odour to that of natural musk and Musk Ambrette®, as its name implies, has an odour reminiscent of ambrette seeds. Other nitromusks which were discovered later include Musk Tibetine® (40) and moskene (41). Musk Xylene® and Musk Ketone® are prepared from *m*-xylene through initial *t*-butylation (Scheme 4.43). Nitration of the *t*-butyldimethylbenzene thus produced gives Musk Xylene®, and acetylation followed by nitration gives Musk Ketone®. The other nitromusks are prepared by similar combinations of classic aromatic reactions.

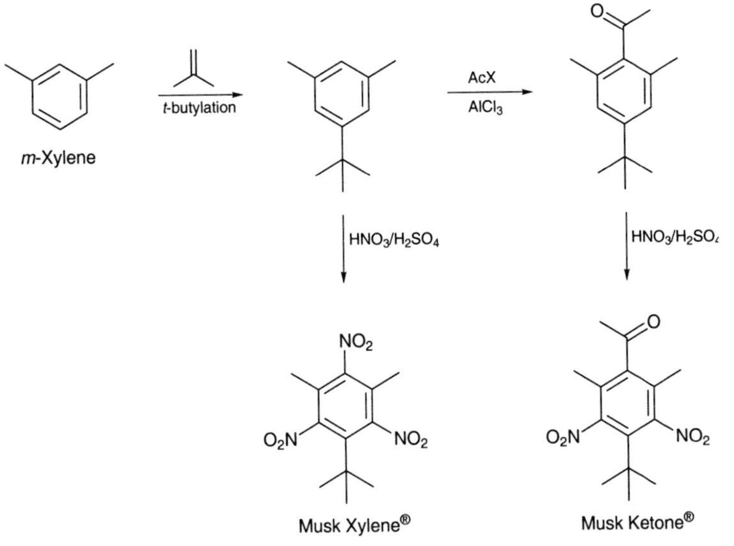

m-Xylene

t-butylation

AcX

$AlCl_3$

HNO_3/H_2SO_4

HNO_3/H_2SO_4

NO_2

O_2N

NO_2

O_2N

NO_2

Musk Xylene®

Musk Ketone®

Scheme 4.43

The nitromusks became the main contributors of musk notes in perfumery and maintained that position until the middle of the twentieth century. However, nitromusks suffer from a number of disadvantages and the discovery of the polycyclic musks in the middle of the twentieth century led to their demise. The preparation of nitromusks can be hazardous (after all, they are related to explosives such as TNT) and, although the final products are not explosive, some of the intermediates and reagents are not free of hazard. Moreover, some nitromusks have been found to be phototoxic; that is, when exposed to sunlight they produce species which can cause allergic reactions on skin. The combined effect of all of these problems, coupled with the ready availability of good alternatives, has made the nitromusk family obsolete.

Polycyclic Musks

In the 1950s, a new class of musks was discovered, the polycyclic musks. These materials mostly have indane, tetralin or isochroman structures, heavily substituted by methyl or other small alkyl groups. Typical members of this family are Traseolide® (42), Phantolide® (43) and Celestolide®/Crysolide® (44).

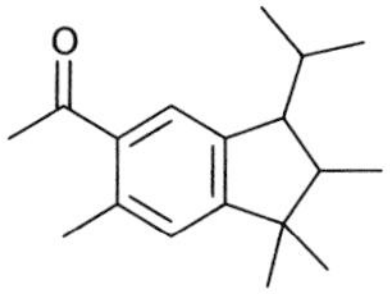

Traseolide®
(42)

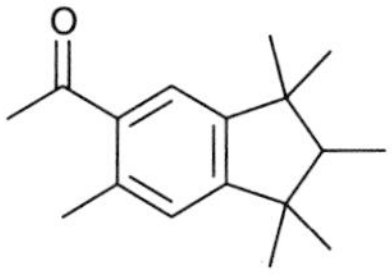

Phantolide®
(43)

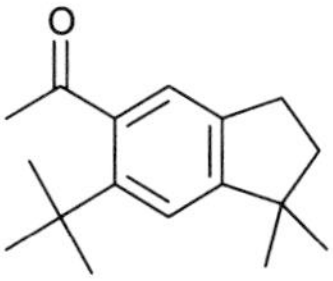

Celestolide®
or
Crysolide®
(44)

Currently, the most important polycyclic musks, in commercial terms, are 6-acetyl-1,1,2,4,4,7-hexamethyltetralin and 4,6,6,7,8,8-hexamethyl-1,3,4,6,7,8-hexahydrocyclopenta[*g*]isochromene. These two compounds are each known by a variety of tradenames, depending on the manufacturer. The commonest tradenames for the tetralin are Fixolide®, Tetralide® and Tonalid® and for the isochroman, Abbalide® and Galaxolide®. Both fragrance materials have been of major importance to the perfume industry throughout the second half of the twentieth century and so it is worthwhile looking in more detail at the chemistry of their syntheses. The synthesis of Galaxolide® is shown in Scheme 4.44.

Cumene is prepared on a large scale from benzene and propylene as an intermediate in the synthesis of acetone and phenol. This makes it an inexpensive and readily available starting material for the production of Galaxolide®. Three further electrophilic addition reactions complete the synthesis. Firstly, isoamylene is added, to form pentamethylindane, to which propylene oxide is added. Finally, treatment with formaldehyde leads, *via* the hemiacetal, to the isochroman.

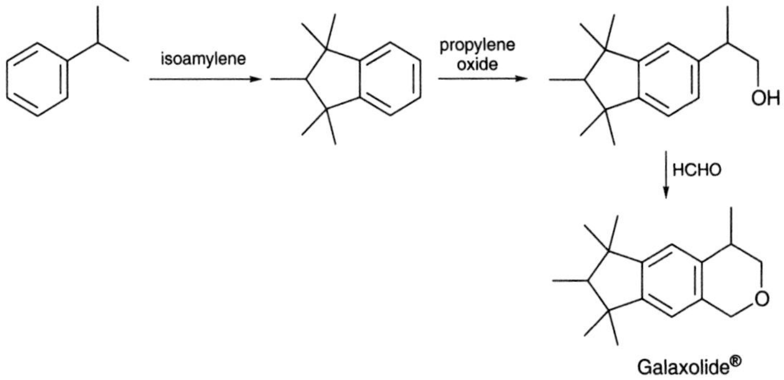

isoamylene
propylene oxide
HCHO
OH
Galaxolide®

Scheme 4.44

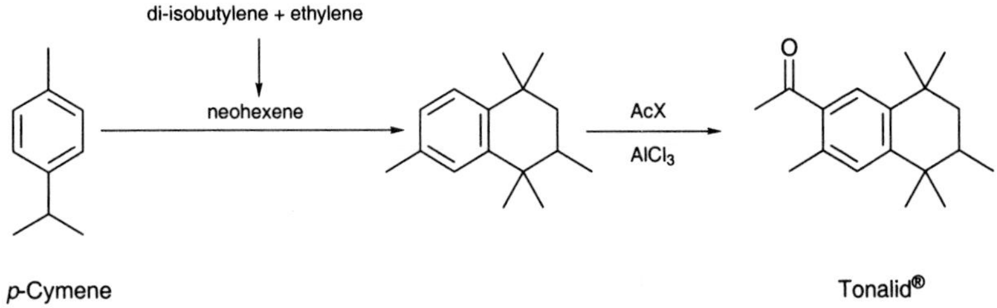

di-isobutylene + ethylene
neohexene
AcX
$AlCl_3$
p-Cymene
Tonalid®

Scheme 4.45

The synthesis of Tonalid® is shown in Sheme 4.45. The starting materials, as for Galaxolide®, are all inexpensive and readily available chemicals. *p*-Cymene is a terpene and is found as a by-product in many processes that involve heating of terpenes, since it lies at the bottom of a potential energy well. Further details are given in the section on terpenes. Dimerization of butylene gives di-isobutylene, which undergoes olefin metathesis with ethylene to give neohexene. The Friedel–Crafts reaction between *p*-cymene and neohexene gives hexamethyltetralin, which can then be acetylated to produce Tonalid®. The cycloalkylation stage in this sequence is interesting and more detail of the mechanism is shown in Scheme 4.46. As written in Scheme 4.45, the reaction requires two moles of neohexene per mole of *p*-cymene. Half of the olefin serves as a hydride abstraction agent (*i.e.* an oxidant) and the other half as the alkylating species. Protonation of the olefin generates a carbocation. This abstracts a hydride ion from the *p*-cymene to give the more thermodynamically stable *p*-cymyl cation. The latter then adds to a second molecule of neohexene. This addition occurs with a concomitant methyl shift, so that the next cation to be formed is tertiary rather than secondary and hence somewhat more stable. This product cation adds to the aromatic ring to give the hexamethyltetralin.

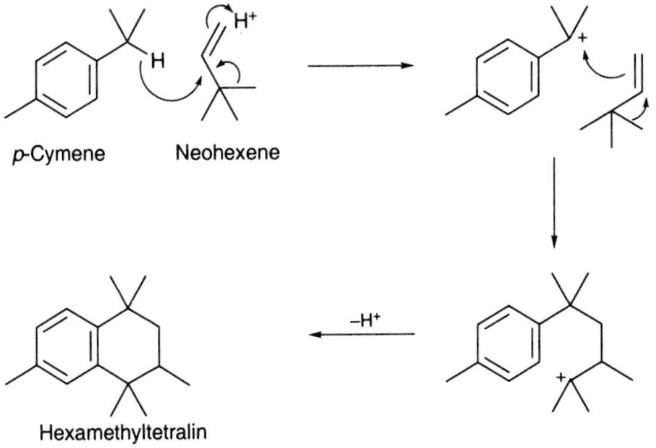

p-Cymene Neohexene

H^+

$-H^+$

Hexamethyltetralin

Scheme 4.46

Neohexene is by far the more expensive of the two starting materials, and so the process shown in Scheme 4.45 is unsatisfactory, since it consumes twice as much of this reagent as is desirable. One way of overcoming this problem is to add a sacrificial oxidant that is less expensive than neohexene. For example, *t*-butyl chloride, under the influence of aluminium chloride, loses its chlorine to produce the *t*-butyl cation, which can abstract the hydride ion from *p*-cymene and thus save one molar equivalent of neohexene.

An alternative solution is shown in Scheme 4.47, the Quest synthesis of a musk mixture known under the trade name Extralide®. In this synthesis, no oxidation is necessary in the cycloalkylation stage because the appropriate carbon atom of the *p*-cymenyl structure is already at the correct oxidation level, since it carries a hydroxyl group. This is achieved by dehydrogenation of α-terpineol to give *p*-cymenol. Reaction of this with 2,3-dimethylbut-2-ene, as an alternative to neohexene, gives a mixture of hexamethyltetralin and isopropyltetramethylindane, acylation of which gives a mixture of two musks. This mixture performs almost identically with pure Tonalid® in perfumes and has the advantage of a lower melting point, which makes dissolution in perfumes easier.

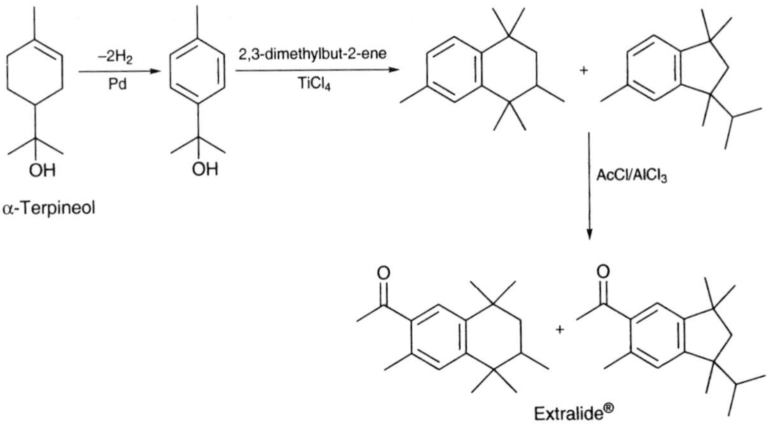

−2H2
Pd
2,3-dimethylbut-2-ene
TiCl4
AcCl/AlCl3
α-Terpineol
Extralide®

Scheme 4.47

Macrocyclic Musks

The problem in the synthesis of macrocyclic musks lies with entropy. The simplest approach to building a large ring is to make a long chain with functionality at each end such that the two ends of the chain can react to close the ring through the formation of a new carbon–carbon bond. However, entropy dictates that the likelihood of the two ends of the chain meeting is lower than that of one end reacting with the end of another chain, repetition of which leads to polymerization. Thus, for example, if 15-hydroxypentadecanoic acid is subjected to esterification conditions, the product is a polyester rather than the musk cyclopentadecanolide. In the 1930s, Stoll solved this problem through the use of extremely high dilution, which reduces the chance of intermolecular reaction and therefore gives the functional groups at each end of the reagent molecule time to find each other and undergo intramolecular reaction. The high dilution principle works well on a laboratory scale, but it is not satisfactory as a manufacturing process because of the poor reactor utilization and the need to recover and recycle very large volumes of solvent. Both of these features lead to unacceptably high process costs.

Many methods have been developed in an attempt to overcome the problem posed by entropy in the formation of large rings. Inclusion of unsaturation in the substrate reduces the number of degrees of freedom in it and consequently increases the possibility of the two ends of the chain meeting. In the acyloin reaction, electrostatic attraction is used to restrict movement of the alkyl chain. A dicarboxylic acid is added to a suspension of sodium (or another alkali metal) particles in an inert solvent. The acid functions react with the metal to form carboxylate anions which are held to the positive surface of the metal by electrostatic attraction. This means that the chain becomes a loop, 'fastened' to the metal particle at each end until the two carboxylate groups approach close enough to allow the acyloin reaction to take place between them.

Another method was developed in the 1930s by Carothers. He began by preparing polyesters from hydroxy acids or from mixtures of dicarboxylic acids and diols. He then depolymerized these to give monocyclic lactones. The lactones are much more volatile than the oligomers and polymers in the mixture and so can be separated from it by distillation. Obviously, this steady removal of the lactones also helps to force the equilibrium in the desired direction. The dicarboxylic acids are available by oxidation of unsaturated fatty acids or cyclic olefins or ketones. For example, ozonolysis of erucic acid, the major fatty acid of oilseed rape, gives brassylic acid (tridecanedioic acid). His first method of depolymerization was to heat the polymer to a high temperature and allow the chains to bite back on themselves, and the volatile lactones were removed by distillation from the mixture. He later developed the useful technique of using high-boiling alcoholic solvents to achieve the depolymerization.

The main alcohol used was glycerol. Glycerol serves two purposes in the depolymerization. Firstly, it provides hydroxyl groups to help keep the interesterification equilibrium reactions in progress. Secondly, the boiling point of glycerol is in the same range as those of lactones in the 15–18 carbon range. By maintaining the system under reflux of glycerol, removal of the lactones by distillation is made more efficient. Furthermore, the lactones are only poorly soluble in liquid glycerol and the distillate readily separates into two phases, making removal of the lactones easy through use of a Dean–Stark trap. Calcium oxide or hydroxide is usually employed as the catalyst in these reactions. Unfortunately, glycerol is not very stable under the conditions and much of it is lost, adding to costs in terms of both materials used and waste disposal. Ethylene brassylate is prepared by this type of process from brassylic acid and ethylene glycol. The low cost of the starting materials and the simplicity of the process make this the least expensive of macrocylic musks at present, and therefore the one which is used in the highest tonnage.

The second largest macrocylic musk, in tonnage terms, is cyclopentadecanolide. Its synthesis is shown in Scheme 4.48. The key starting material is methyl undecylenate, which is obtained from castor oil by pyrolysis and esterification. Exposure of tetrahydrofuran to a radical initiator generates the radical at a carbon next to the oxygen atom. This radical adds to the terminal double bond of methyl undecylenate and the radical produced then abstracts a hydrogen atom from another tetrahydrofuran and so propagates the radical chain reaction. The methyl 11-(2′-tetrahydrofuryl)undecanoate thus produced is subjected to elimination, hydrogenation and hydrolysis to give 15-hydroxypentadecanoic acid. Polymerization–depolymerization gives cyclopentadecanolide.

Tetrahydrofuran + Methyl undecylenate $\xrightarrow{R^\bullet}$ (2-tetrahydrofuryl)undecanoate (CO_2Me)

1) heat
2) H_2/catalyst
3) H_3O^+

HO / HO_2C (15-hydroxypentadecanoic acid) $\xrightarrow{-H_2O}$ Cyclopentadecanolide

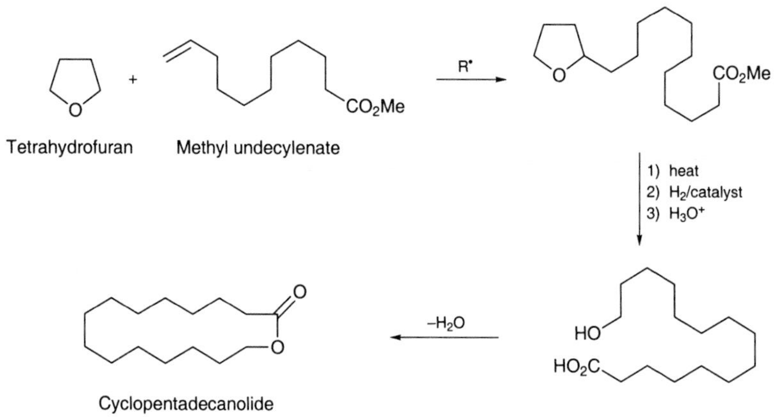

Scheme 4.48

In 1959, Wilke discovered that butadiene could be trimerized round a metal template to give cyclododecatriene. This could be converted into the mono-olefin, the ketone, the alcohol, *etc*., by obvious means. The ready availability of an inexpensive supply of cyclododecane derivatives set in train a new direction in musk research. The 12-carbon ring could be broken open, with or without addition of a side chain, to provide new linear precursors for macrocyclization. Furthermore, the ring could be enlarged by fusing a second ring to it, breaking the bridgehead bond to produce a larger ring. This latter option offers an elegant means of overcoming the entropic problem of macrocyclization. An example of this type of approach is shown in Scheme 4.49.

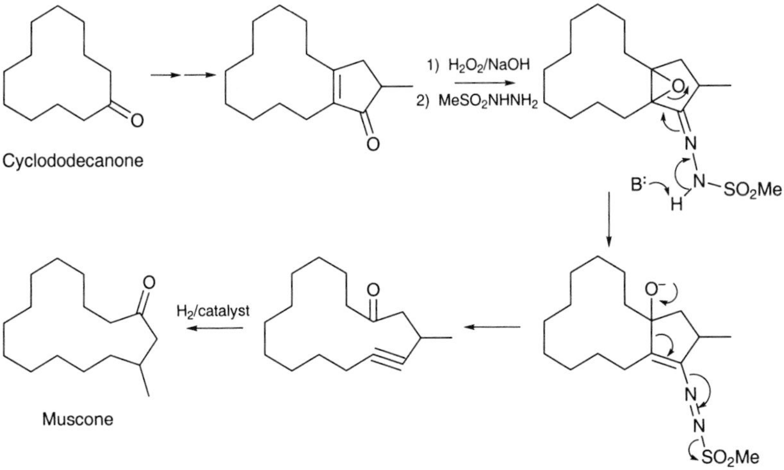

Cyclododecanone

1) H_2O_2/NaOH
2) $MeSO_2NHNH_2$

B:

SO_2Me

H_2/catalyst

Muscone

Scheme 4.49

Eschenmoser's synthesis of muscone (Scheme 4.49) uses cyclododecanone as the starting material and employs the Eschenmoser fragmentation reaction in the key bridgehead-breaking step. Firstly, a methacrylate unit is fused to the 12-membered ring using conventional anionic chemistry. This leads to the bicyclic ketone which can be converted into the epoxide with alkaline hydrogen peroxide. Addition of methane sulfonylhydrazine gives the hydrazone. When treated with base, the hydrogen attached to the nitrogen atom is lost and the resultant negative charge flows through the molecule to spring open the epoxide ring. The negative charge can then flow back across the bridgehead by a different route and the molecule fragments, losing nitrogen. It is the energy gain in forming free nitrogen that drives this reaction. The acetylenic ketone produced is easily hydrogenated to produce muscone. Routes such as this are very elegant examples of chemical synthesis, but they are multi-step and therefore attract high process costs when operated on an industrial scale. At present, no macrocyclic musks are available in the same cost bracket as the polycyclic musks. Consequently, the fragrance industry still carries out a great deal of research into methods of producing macrocyclic ketones and lactones at lower cost.

PERFUME INGREDIENTS DERIVED FROM BENZENE

One of the most important perfume ingredients made from benzene is

2-phenylethanol. The synthesis of this, and related compounds, is shown in Scheme 4.50. 2-Phenylethanol is a major component of rose oils and is widely used in perfumery for its blending qualities. In tonnage terms, it is one of the most important of all perfumery ingredients. The original production method involved Friedel–Crafts addition of ethylene oxide to benzene, which is fairly efficient despite some addition of ethylene oxide to the product that results in a small amount of polyethoxylated derivatives. The major disadvantage of this route is the safety issue of handling ethylene oxide and benzene. Both reagents can be handled safely, but the engineering required to do so adds to the process costs.

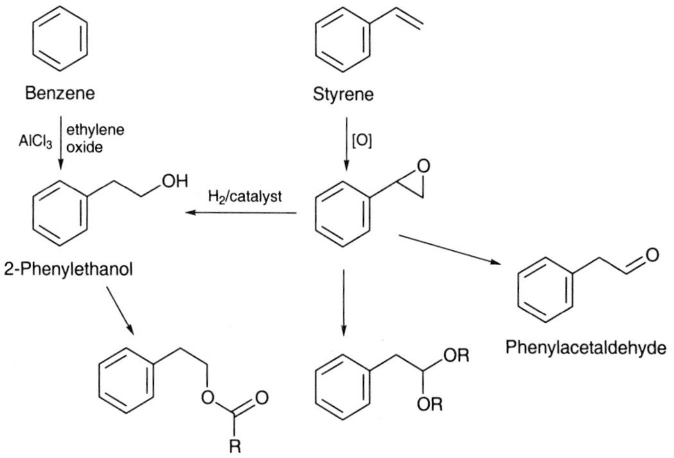

Benzene

Styrene

$AlCl_3$ ethylene oxide

[O]

OH

H_2/catalyst

O

2-Phenylethanol

O

Phenylacetaldehyde

O O R

OR OR

Scheme 4.50

Another possibility is to hydrogenate styrene oxide (the production of styrene from benzene is described below). This route gives a very high-quality product and the intermediate, styrene oxide, can also be used to produce other fragrance materials. Rearrangement of the epoxide gives phenylacetaldehyde, which is a potent green note. Phenylacetaldehyde is responsible for the characteristic green top note of narcissus and is used to create a narcissus-like effect in floral perfumes. Addition of alcohols to styrene oxide gives the corresponding acetals of phenylacetaldehyde. The most important of these is the dimethyl acetal, known in the industry by the acronym PADMA. This material has much of the green character of the aldehyde, but is chemically more stable. Many esters of 2-phenylethanol are used in perfumery, the acetate, isobutyrate and phenylacetate in particular. These are prepared by esterification of the alcohol.

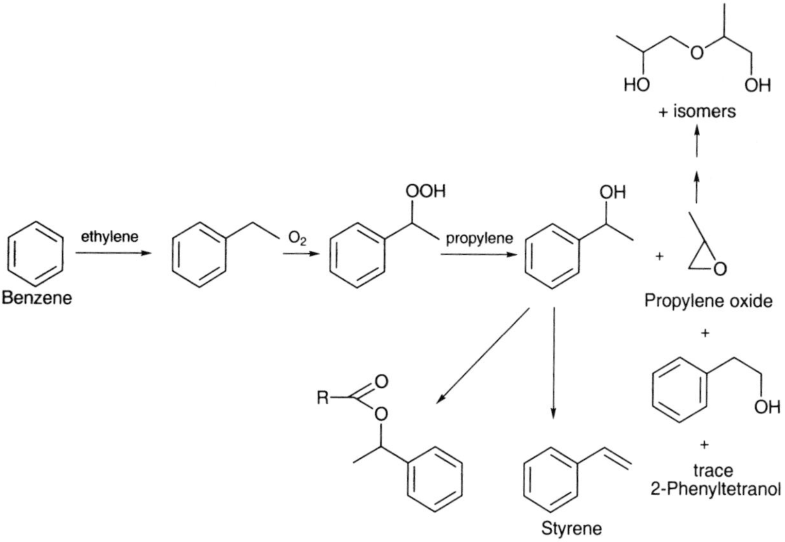

Benzene ethylene O2 OOH propylene OH + isomers Propylene oxide + OH + trace 2-Phenyltetranol Styrene

Scheme 4.51

Scheme 4.51 shows the process known to the bulk chemicals industry as SMPO, the styrene monomer–propylene oxide process. Styrene is used in polymers, and propylene oxide derivatives have a wide variety of uses, including as surfactants and in anti-freeze. For the bulk industry, the process is as follows. Addition of ethylene to benzene gives ethylbenzene, which undergoes air oxidation to give the hydroperoxide. Reaction of this with propylene, in the presence of a suitable catalyst, gives styrallyl alcohol and propylene oxide. Styrallyl alcohol is readily dehydrated to styrene.

Perfumery interest in this process is four-fold. Firstly, one of the major products, styrene, is a starting material for perfume ingredients, as described above. Secondly, the other major product, propylene oxide, is also a precursor for a number of fragrance materials and for dipropylene glycol, one of the major solvents used in perfumes. Thirdly, the intermediate, styrallyl alcohol, is the starting material for a number of esters used in perfumery, the acetate in particular. Fourthly, but by no means least important, the crude styrallyl alcohol contains a small trace of 2-phenylethanol. Since the SMPO process is run on a scale far beyond that of the perfume industry, what is a small trace to the polymer business represents a very significant proportion of the world requirement for 2-phenylethanol as a perfume ingredient. The problem is that of odour quality.

It is extremely difficult to purify this by-product 2-phenylethanol to the odour quality of that produced by either of the above routes. However, most of the companies, such as ARCO in the USA and Sumitomo in Japan, who run the SMPO process can produce 2-phenylethanol of a quality which can be used in perfumery (often in collaboration with a perfumery company). The amount of 2-phenylethanol available from this route is dictated by the demand for styrene and propylene oxide, the market value is dictated largely by material from the other two routes and all three run in economic balance.

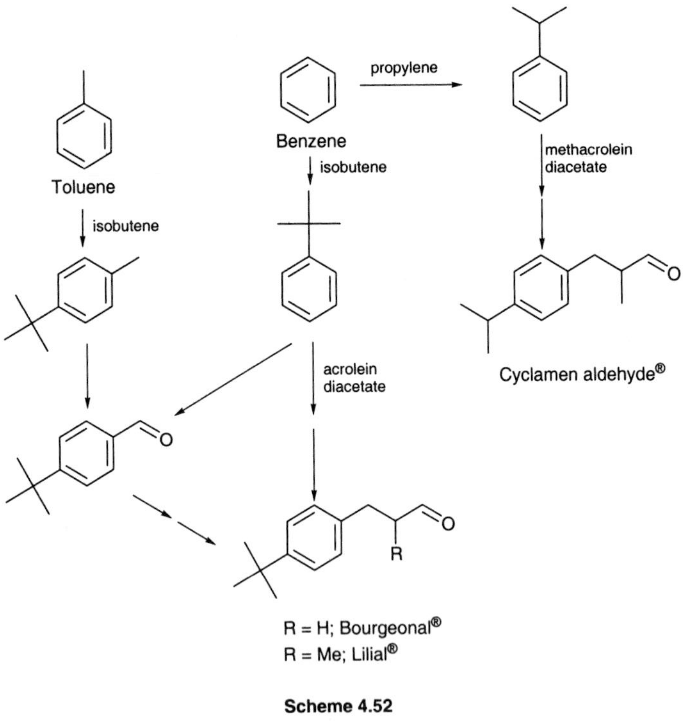

Toluene
isobutene
Benzene
isobutene
propylene
methacrolein diacetate
acrolein diacetate
Cyclamen aldehyde®
R = H; Bourgeonal®
R = Me; Lilial®

Scheme 4.52

Scheme 4.52 shows the preparation of the hydrocinnamic aldehydes, another family of materials derived from benzene and which possess fresh, white-floral notes reminiscent of muguet (lily of the valley) and cyclamen. One of these, Lilial® or Lilistralis®, can also be prepared from toluene, and that route is also described in this section for comparison. Addition of isobutene or propylene to benzene gives *t*-butylbenzene and cumene, respectively. Addition of acrolein or methacrolein diacetates to these gives, after hydrolysis of the intermediate enol ester, the corresponding hydrocinnamaldehydes. It is also possible to prepare the hydrocinnamaldehydes from the substituted toluene by oxidation to the corresponding substituted benzaldehyde, followed by aldol condensation and hydrogenation. The oxidation of the substituted toluenes to the corresponding substituted benzaldehydes can be carried out either by electrochemical oxidation or by chlorination–hydrolysis. This, however, is not the case for oxidation of *p*-cymene (*p*-isopropyltoluene), since in this case the isopropyl group is more reactive than the methyl group.

The substituted benzaldehydes can also be prepared by formylation of the corresponding alkyl benzenes. In Scheme 4.52, all of the possibilities are shown for the preparation of Lilial®. The most important route commercially is that which proceeds from toluene through electrochemical oxidation of *t*-butyltoluene followed by aldol reaction and hydrogenation. Lilial® is an intermediate for a herbicide, and so some producers gain the advantage of scale in their production costs by making both products. Cumene is produced on a vast scale as a precursor for acetone and phenol, which makes it the most sensible starting material for cyclamen aldehyde. As with so many perfumery materials, this family of aldehydes demonstrates how various synthetic routes are possible; the ones chosen depend on a fine balance of technical, economic and strategic factors.

PERFUME INGREDIENTS DERIVED FROM TOLUENE

The alkylation of toluene to give hydrocinnamic derivatives is discussed in the previous section, alongside the same reaction of benzene. The tree of products in this section all derive from the oxidation of toluene, as shown in Scheme 4.53.

Air oxidation of toluene gives predominantly benzoic acid. This is used in perfumery for the preparation of benzoate esters, benzophenone and various other compounds, but this use is dwarfed by the other industrial uses of benzoic acid. It is used in many different ways, for instance, as a precursor for nylon monomers. Crude benzoic acid contains a small amount of benzaldehyde, which is easily extracted from it. In view of the huge volume of benzoic acid produced, the volume of benzaldehyde recovered makes a substantial contribution to that used by the perfumery industry.

Chlorination of toluene under radical conditions (either through the

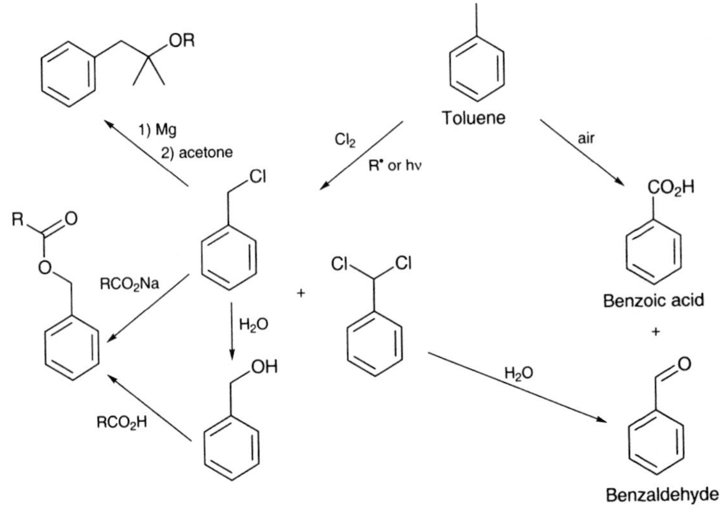

Scheme 4.53

use of an initiator or by photolysis) gives a mixture of mono-, di- and trichlorotoluene. In practice in industry, the reaction is run with an excess of toluene present which means that benzyl chloride is the major product. A little benzal chloride is produced and can be separated and hydrolysed to give benzaldehyde. The major use of benzyl chloride is in the production of benzyl alcohol and its esters; the alcohol is produced by hydrolysis of the chloride. The esters can be prepared by esterification of the alcohol, but it is better economically to prepare them directly from the chloride by reaction with a salt of the corresponding acid. By far the most important of this group of products is benzyl acetate, the major component of jasmine oils. Grignard addition of benzyl chloride to acetone leads to a family of ingredients based on dimethylbenzylcarbinol. The most important member of the family is the acetate, known by the acronym DMBCA.

Although benzaldehyde has an odour which is very characteristic of almonds, it is the chemicals derived from it, rather than its own odour, that make it an important material to the perfumery industry. The major benzaldehyde derivatives are shown in Scheme 4.54. The most important are shown on the left of the scheme.

Claisen ester condensation gives cinnamic acid and its esters, the most important of which is methyl cinnamate, followed by benzyl cinnamate. Aldol condensation of benzaldehyde with other aldehydes

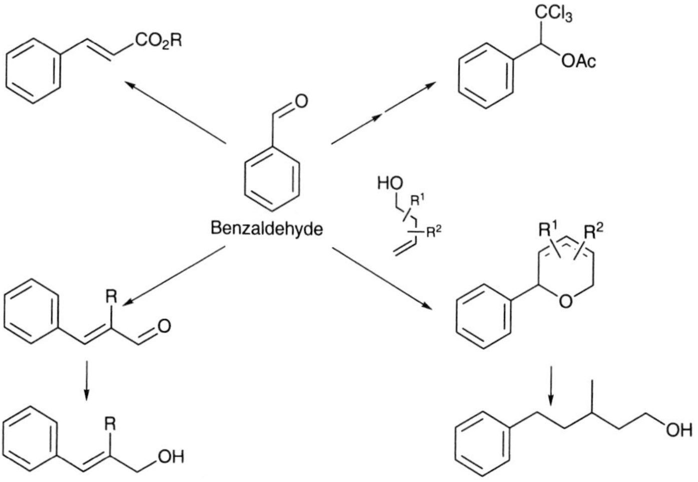

R = H; cinnamaldehyde
R = *n*-amyl; amylcinnamic aldehyde
R = *n*-hexyl; hexylcinnamic aldehyde

Scheme 4.54

gives the series of α-substituted cinnamaldehydes. The lowest member of the series, cinnamaldehyde, is used to some extent in fragrances, but its main use is as the starting material for the corresponding alcohol, cinnamyl alcohol. This is an important component of spicy perfumes in which a cinnamon note is required. Its esters, the acetate in particular, are also used for their odours. Much more important are the higher members of the series, amylcinnamic aldehyde (ACA) and hexylcinnamic aldehyde (HCA). They possess odours reminiscent of the fatty background note of jasmine, although neither are found in jasmine oils. Most synthetic jasmine perfumes use one or both of these compounds as the foundation on which the fragrance is built. They are inexpensive materials and so can be used in large proportions in perfume formulae. They are also very fibre-substantive materials and are therefore of great importance in laundry products, such as detergents and fabric conditioners.

Addition of chloroform to benzaldehyde followed by esterification with acetic anhydride gives the trichloro derivative known as rose crystals or, more commonly by the misnomer, rose acetone. Such misnomers are not uncommon with older fragrance materials. Some are accidental, but others were probably intended to deceive competitors in the days before analytical chemistry progressed to the stage where such deception is easy to uncover.

Benzaldehyde undergoes the Prins reaction with homoallylic alcohols to give a variety of perfume ingredients, mostly with green, herbaceous odours. These products have very intense odours and so are used at relatively low levels in fragrances. The product obtained from the Prins reaction of benzaldehyde and 3-methylbut-3-ene-l-ol (isoprenol) can be hydrogenated to give 3-methyl-5-phenylpentan-l-ol, the pyran ring being broken open by hydrogenolysis of the benzylic ether bond during the hydrogenation. This material is known under various tradenames, such as Mefrosol® and Phenoxanol®. It has a very pleasant, fresh, white-floral odour.

PERFUME INGREDIENTS DERIVED FROM PHENOL

Phenol is a material of major commercial importance. One of its earliest uses was as a disinfectant (carbolic acid). Earlier in the twentieth century, it became important as a feedstock for resins such as Bakelite®, and in the latter part of the century it also became very important as a precursor for caprolactone and caprolactam and hence polyester and polyamide manufacture. The two major methods for phenol production nowadays are by the catalytic oxidation of benzoic acid and catalytic decomposition of cumene hydroperoxide (Scheme 4.55).

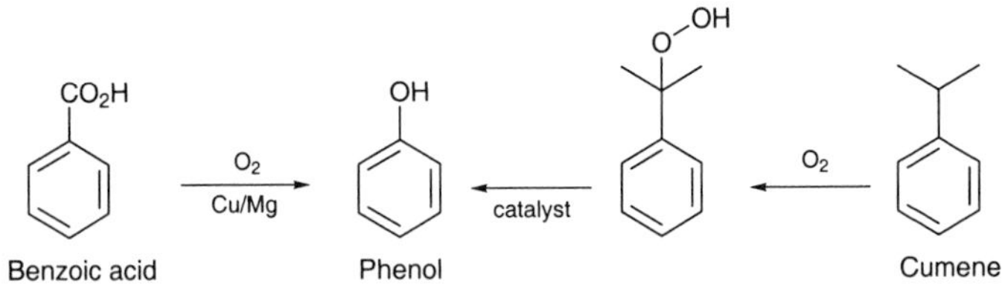

CO_2H OH O OH O_2 Cu/Mg catalyst O_2

Benzoic acid Phenol Cumene

Scheme 4.55

Diphenyl oxide, prepared from phenol, is important in rose and other floral fragrances. The addition of ethylene oxide to phenol gives phenoxyethanol and hence its esters, the most important of which is the isobutyrate (Scheme 4.56). Etherification gives materials such as anisole (methyl phenyl ether), estragole [3-(*p*-methoxyphenyl)prop-1-ene, a constituent of tarragon] and anethole [1-(*p*-methoxyphenyl)-prop-l-ene, which occurs in and is strongly characteristic of aniseed].

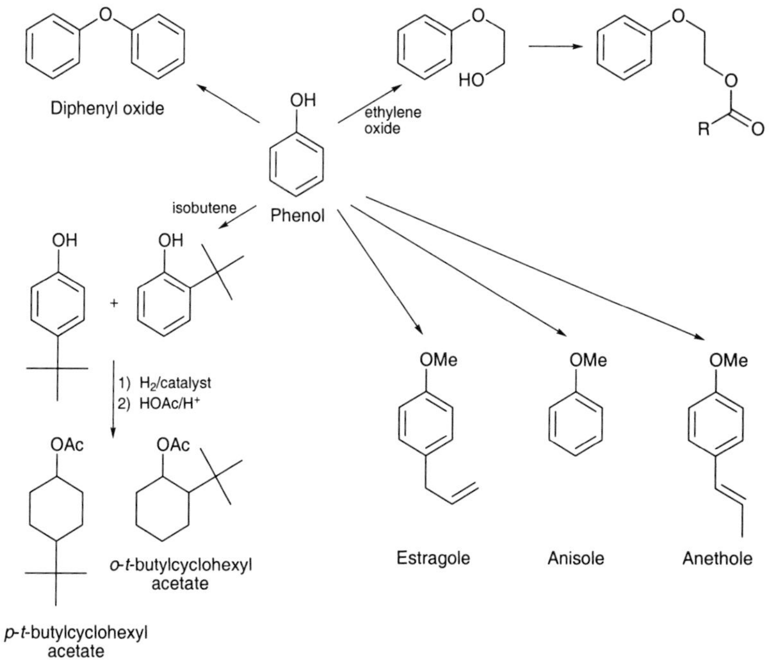

Diphenyl oxide
OH
ethylene oxide
HO
R
O
isobutene
Phenol
OH
OH
+
1) H_2/catalyst
2) HOAc/H^+
OAc
OAc
OMe
OMe
OMe
o-t-butylcyclohexyl acetate
Estragole
Anisole
Anethole
p-t-butylcyclohexyl acetate

Scheme 4.56

Anethole and estragole occur in sulfate turpentine. Distillation from this source provides much of the anethole and estragole required, the shortfall in supply being made up by material synthesized from anisole. The addition of isobutylene to phenol gives a mixture of *o*- and *p-t*-butylphenols, which can be separated and hydrogenated to the corresponding cyclohexanols and then esterified to produce the acetates. The acetates are both very important fragrance ingredients and are used in large quantities. They are known as OTBCHA and PTBCHA, acronyms for *o*- and *p-t*-butylcyclohexyl acetate, respectively. OTBCHA is also known as Ortholate® and has an apple odour with some fresh woody notes. PTBCHA has a warm, sweet, fruity, woody character. Both are mixtures of *cis*- and *trans*-isomers. The *cis*-isomer of PTBCHA is stronger than the *trans*- and has a desirable jasminic–floral character. In consequence, PTBCHA is sold in regular and high *cis*- grades. It is possible to separate the isomers completely, but the cost of doing so is prohibitive.

Addition of a one carbon unit to phenol in a Friedel–Crafts type of reaction gives rise to a family of perfume ingredients of great importance, as shown in Scheme 4.57. Carboxylation gives salicylic acid. Acetylation of the phenolic group of salicylic acid gives aspirin and thus the acid is an important commodity chemical for the pharmaceutical industry. The esters of salicylic acid are important to the fragrance industry. Methyl salicylate is the major component of oil of wintergreen. Sportsmen and women readily recognize its odour, since it is characteristic of liniments prepared from that oil. The most important salicylates to perfumery are the amyl, hexyl and benzyl derivatives, which are used in very significant quantities. These have persistent, floral, herbaceous odours and make excellent blenders and fixatives for floral perfumes. The odour which we have all come to recognize as characteristic of sun-tan lotion is largely that of these higher salicylate esters.

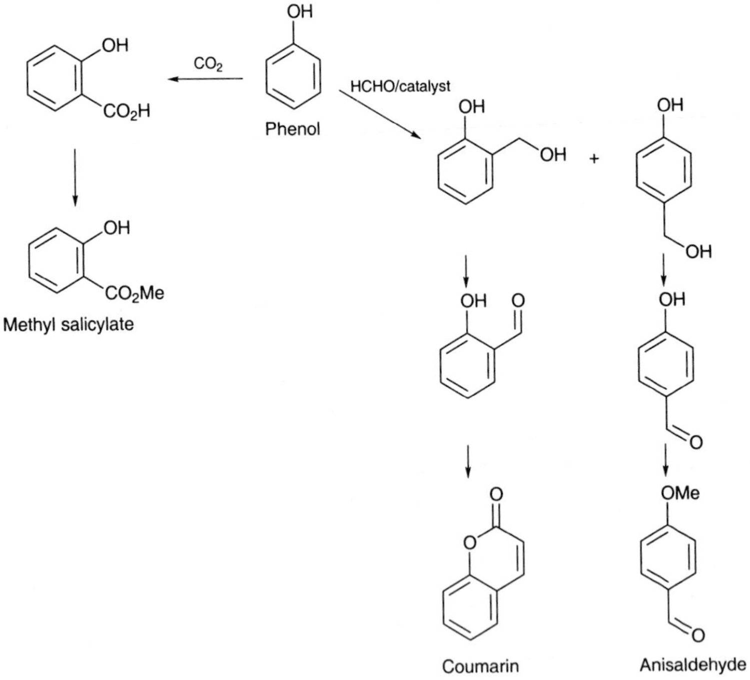

Scheme 4.57

Addition of formaldehyde to phenol normally produces a resin. However, under controlled, catalytic conditions, it is possible to obtain the hydroxymethyl derivatives in high yield. *o*-Hydroxymethyl-phenol is known as saligenin and can be oxidized to salicylaldehyde. Similarly, the *p*-isomer can be oxidized to *p*-hydroxybenzaldehyde. Methylation of the latter gives anisaldehyde, which has a hawthorn odour, and Perkin condensation of the former leads to coumarin. Coumarin is present in newly mown hay and its sweet, hay-like note is used widely in perfumery. Thus, one reaction leads to the precursors for two important fragrance materials. The ratio of *o*- to *p*-substitution can be controlled to some extent by the choice of catalyst and conditions, and so the reaction product mixture can be adjusted to suit the demand for the two end products. (Originally, salicylaldehyde was prepared from phenol using chloroform in the Reimer–Tiemann reaction.)

Oxidation of phenol leads to *o*-dihydroxybenzene (catechol, Scheme 4.58). This oxidation can be carried out in a number of ways, but the most important commercially is to use hydrogen peroxide as the oxidant with iron salts as catalysts. The monomethyl ether of catechol is known as guaiacol and can be prepared from catechol by partial methylation using dimethyl sulfate or an equivalent reagent. Guaiacol is a precursor for the terpenophenol family of sandalwood chemicals, as discussed earlier in the section on terpenes. Formylation of guaiacol gives vanillin (Scheme 4.58), the character-impact component of vanilla. The ethyl analogue can be prepared similarly and is known as ethylvanillin. Formylation of the methylene ether of catechol gives methylenedioxybenzaldehyde, which is commonly known as either piperonal or heliotropin. The former name comes from the corresponding acid, piperonylic acid, which is a degradation product of the pungent principles of pepper. The latter, and more common name, is derived from the fact that the odour of the aldehyde is strongly reminiscent of heliotropes, since it is the major fragrant component of those flowers. The vanillic group of compounds are used to give a heavy and long-lasting sweetness to fragrances, a character which is currently much in vogue with perfumes such as *Tocade* and *Angel*.

Vanillin and ethylvanillin are not particularly stable chemically. This is not surprising, since they possess both an aldehyde and a phenolic group. In functional products, where the pH is not neutral, they undergo a variety of reactions leading to discoloration. For example, inclusion of vanillin in a white soap will, after a matter of days, produce a colour close to that of chocolate.

The marketing phenomenon known as 'trickle-down' is when the odour of a fine fragrance is adapted so that a range of cosmetics, toiletries, soaps and so on can have the same fragrance to produce a line of products. If the fine fragrance contains vanillin, trickle-down

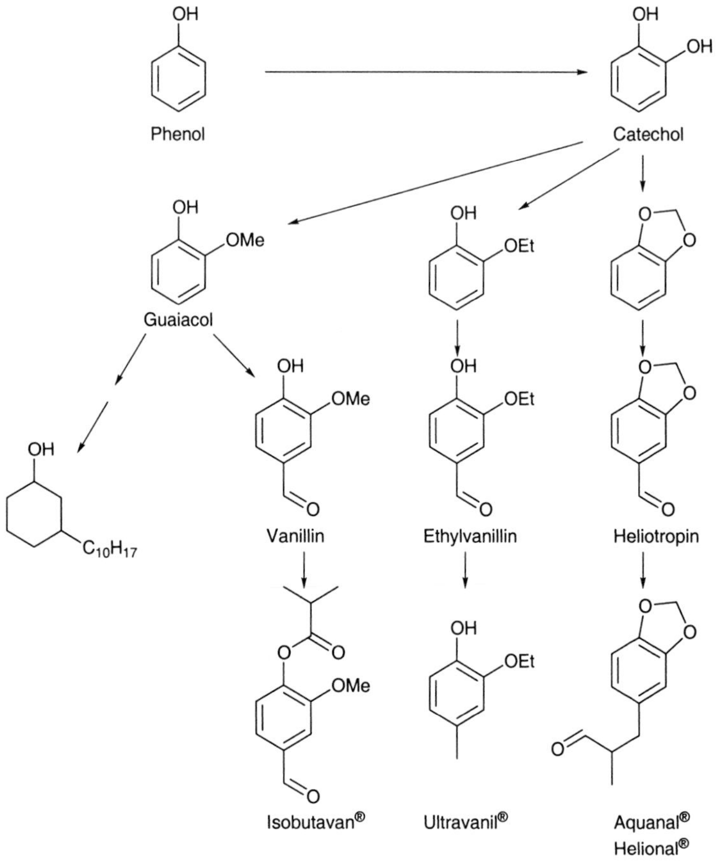

OH OH
OH
Phenol Catechol
OH
OMe
OH
OEt
O
O
Guaiacol
OH
OMe
OH
OEt
O
O
OH
$C_{10}H_{17}$
O
Vanillin
O
Ethylvanillin
O
Heliotropin
O
O
OMe
OH
OEt
O
O
O
O
Isobutavan®
Ultravanil®
Aquanal®
Helional®

Scheme 4.58

is difficult because of the discoloration issue. One task of the fragrance chemist is to overcome such problems and two solutions to this, developed by chemists at Quest, are shown in Scheme 4.58. Protection of the phenolic group of vanillin through the isobutyrate ester gives Isobutavan®. Reduction of the aldehyde group of ethylvanillin to a methyl group gives Ultravanil®. (Interestingly, the corresponding methyl ether, obtained by reduction of vanillin, has a strong smoky odour with no trace of vanillic sweetness.) Both of these compounds provide vanillic notes, but are much more stable in use than vanillin itself.

Aldol reaction between heliotropin and propionaldehyde, followed by hydrogenation, gives the hydrocinnamaldehyde derivative known as

Helional® or Aquanal®. This has a floral note, and is somewhat sweeter than the hydrocinnamaldehydes described in the section on benzene-based materials.

The catechol derived materials can also be approached by a different route, using readily available natural products, as shown in Scheme 4.59. Clove oil is available at moderate cost and in moderate quantity from several tropical countries, such as Indonesia and the Malagasy Republic. Similarly, sassafras oil is available from Brazil. The major component of clove oil is eugenol, 3-(3-methoxy-4-hydroxyphenyl)-prop-l-ene, which can be isomerized to isoeugenol using basic or metallic catalysts. Ozonolysis of isoeugenol gives vanillin. Similarly, heliotropin can be obtained from safrole, 3-(3,4-methylenedioxy-phenyl)prop-1-ene, the major constituent of sassafras oil. The various ethers and esters of these materials shown in Scheme 4.59 are used in perfumery. Eugenol itself is, by far, the most important of these. As well as being present in cloves, eugenol is an important contributor to the odour of carnations and is used in fragrances to that effect. Generally, the catechol route is cheaper than the eugenol–safrole route to vanillin and heliotropin, but local economics may tip the balance in the other direction particularly for heliotropin (which is the more expensive of the two).

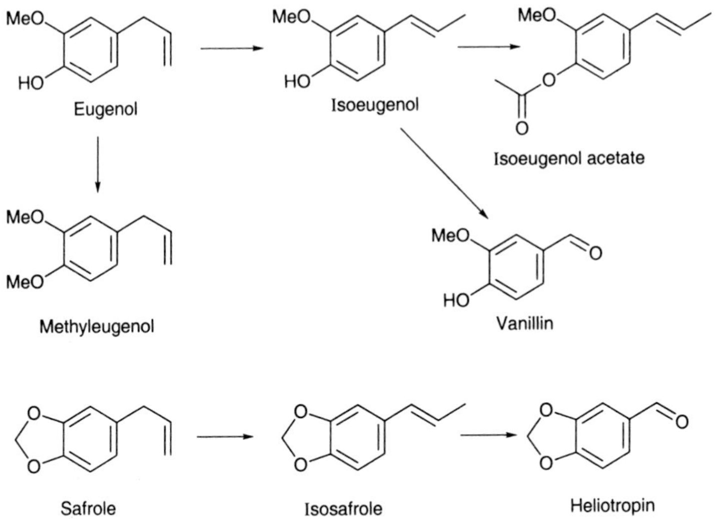

Eugenol Isoeugenol Isoeugenol acetate

Methyleugenol Vanillin

Safrole Isosafrole Heliotropin

Scheme 4.59

PERFUME INGREDIENTS DERIVED FROM NAPHTHALENE

Acetylation of naphthalene gives methyl naphthyl ketone, and sulfonation followed by alkaline fusion gives naphthol (Scheme 4.60). The methyl and ethyl ethers of naphthol are prepared from naphthol by reaction with the corresponding alkyl sulfate under basic conditions. These ethers are usually known by the shorter names of yara and nerolin, respectively. Yara, nerolin and methyl naphthyl ketone possess floral odours and are moderately important perfume ingredients.

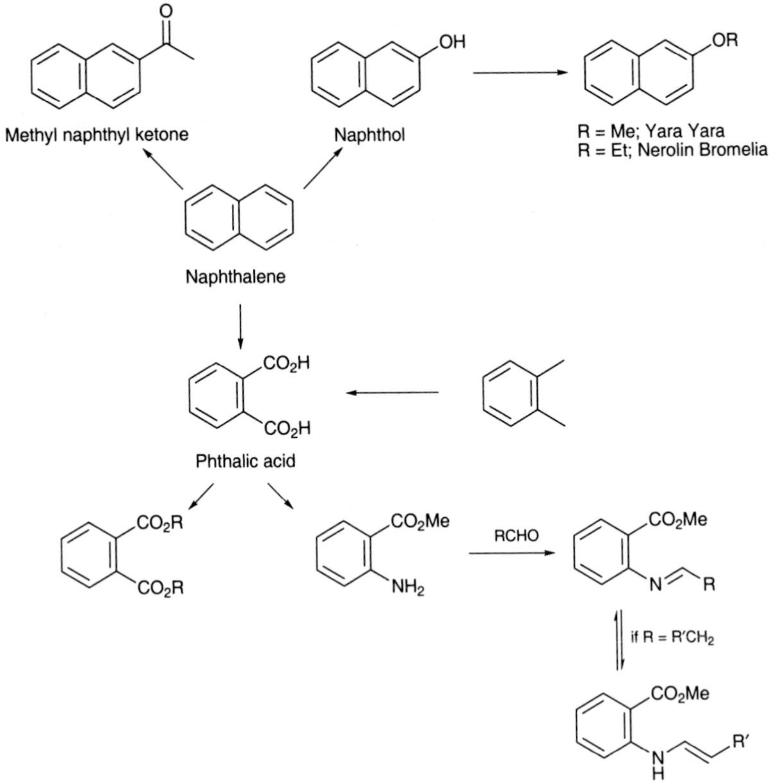

Scheme 4.60

Oxidation of naphthalene gives phthalic acid, which can also be obtained by air oxidation of *o*-xylene. Phthalic acid is an important feedstock for polymers and plasticizers and so, once again, the fragrance industry is 'piggy-backing' on a larger industry to obtain inexpensive ingredients. Higher esters of phthalic acid are used as plasticizers and solvents; the ester of most interest to the fragrance industry is the diethyl ester, which is used as an odourless solvent for fragrances.

Methyl anthranilate is obtained by the Hoffmann rearrangement of phthalimide, which is easily prepared from phthalic anhydride and ammonia. Methyl anthranilate occurs naturally in many flowers and has a very characteristic and intense, sweet smell. When added to an aldehyde, it forms the corresponding Schiff's base. If the aldehyde carries an α-hydrogen atom, the Schiff's base is in equilibrium with the corresponding enamine. The relative proportions of Schiff's base and enamine present depend on the structure of the aldehyde, since the inductive and resonance effects seek to minimize the total free energy of the molecule. Perfumers refer to the products as Schiff's bases irrespective of the exact composition. Indeed, Schiff's bases formed from methyl anthranilate are so important to the industry that the term 'Schiff's base', when used in perfumery, almost invariably implies a Schiff's base of methyl anthranilate. The importance of these Schiff's bases to perfumery results from the fact that they increase both the chemical stability and tenacity of the aldehyde component. The Schiff's bases are less reactive than the free aldehyde and, since the molecular weight is much higher, they are less volatile. Thus, loss of the aldehyde by both chemical reaction and evaporation is slowed down. Moreover, hydrolysis of the Schiff's base releases both the aldehyde and the methyl anthranilate, both of which have intense odours. Thus, use of a Schiff's base generates a long-lived fragrance composition of the two ingredients. Originally, the aldehydes used in these Schiff's bases also had sweet, floral odours and so an harmonious accord was produced. *Giorgio* is an example of a fragrance which relies heavily on this type of accord.

A new fashion was created when Dior introduced *Poison* in which the aldehyde component of the Schiff's base has a strident green character, creating a sharp contrast with the heavy sweetness of the methyl anthranilate. The pale yellow colour of methyl anthranilate is darkened when a Schiff's base is formed from it, because of the hypsochromic shift in the ultraviolet absorbtion maximum that results from the extension of conjugation. This colour has an effect on the colour of any fragrance in which it is incorporated and has to be taken into account if colour might be deleterious in the final product. Many Schiff's bases are available commercially, but some perfumers prefer to formulate their fragrances using aldehydes and free methyl anthranilate and allow the Schiff's base to form spontaneously in the perfume.

PERFUME INGREDIENTS DERIVED FROM ALIPHATIC MATERIALS

A large number of aliphatic fragrance ingredients are used, but few in significant tonnage. This is largely because the materials of use are mostly aldehydes, nitriles and lactones, the majority of which have very intense odours that limit the amount that can be incorporated into a fragrance. A number of volatile esters are also used to give fruity top-notes but, again, these are not used at high levels. The majority of ingredients of this class can be prepared by straightforward synthetic reactions and functional group interconversions, starting from both natural and petrochemical precursors. Some examples are shown below (45–50).

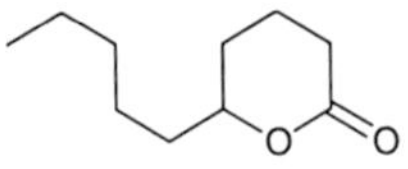

δ-Decalactone
(peach, coconut)

(45)

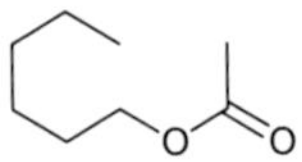

Hexyl acetate
(fruity, pear)

(46)

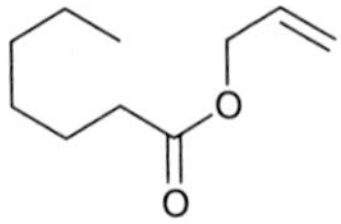

Allyl heptanoate
(sweet, banana)

(47)

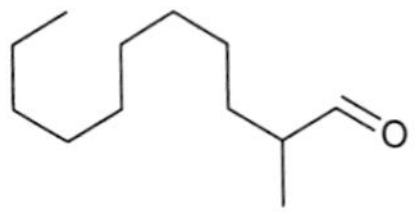

Aldehyde MNA
(fresh, aldehydic)

(48)

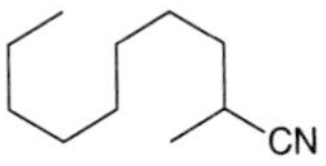

Frutonile®
(soft, floral, fruity)

(49)

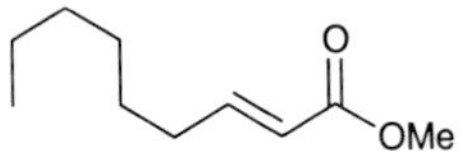

Beauvertate®
(vegetable, earthy)

(50)

The aliphatic fragrance materials of natural origin are mostly derived from fatty acids and related materials. As a result of their biosynthetic pathway (see Chapter 3), those formed directly all have an even number of carbon atoms in the chain. Any with an odd number of carbons in the chain are likely to be breakdown products. Fatty acids are useful precursors for aliphatic fragrance materials, but the same observations about the number of carbon atoms in the chain, of course, also apply. Thus, for example, octanal can be prepared by oxidation of octanol, which can be obtained from coconut oil. Similarly, materials with 10-, 12- or 14-carbon chains can be obtained from other fats. One example of a fatty acid being used in fragrance is myristic acid, the isopropyl ester of which is employed as a solvent. Once a fatty acid or alcohol is available, appropriate oxidation and reduction reactions open routes to the other members of the series. To take octanol as an example again, it can be dehydrogenated to give octanal, which is used both as an ingredient in its own right and as a precursor for other materials such as HCA (see page 107).

Two major reactions are used to break longer chains and give odd-numbered fragments as fragrance building blocks. The first is oxidative cleavage of a double bond in an unsaturated fatty acid. Ozonolysis is a convenient method for doing this and is used to provide pelargonic acid (heptanoic acid) and azelaic acid (nonanedioic acid) from oleic acid, the major constituent of olive oil, and brassylic acid (tridecanedioic acid) from the erucic acid (13-docosenoic acid) of rapeseed oil.

The other reaction is pyrolytic cleavage (*via* a *retro*-ene reaction) of a fatty acid containing a homoallylic alcohol in the chain. This is used to produce two important feedstocks, heptanal and undecylenic acid, from ricinoleic acid, the major fatty acid component of castor oil. The mechanism of this reaction is shown in Scheme 4.61.

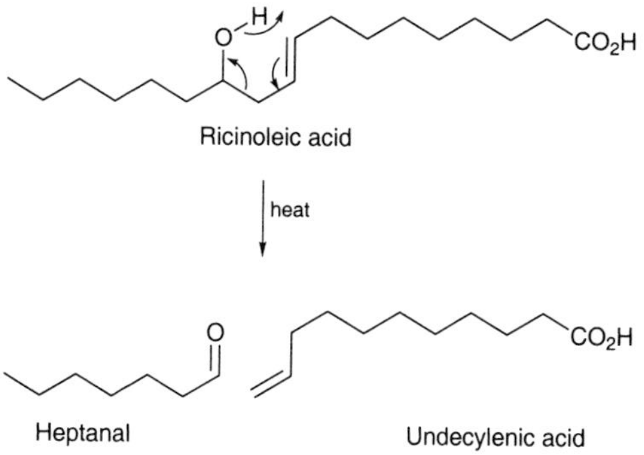

CO_2H

Ricinoleic acid

heat

CO_2H

Heptanal

Undecylenic acid

Scheme 4.61

Undecylenic acid is important as a starting material for a number of fragrance materials, some of which are shown in Scheme 4.62. The two aldehydes are obtained by simple functional group manipulation. Both are known by trivial, and somewhat misleading, names in the industry; 10-undecenal as Aldehyde C11 undecylenic or simply Aldehyde C11, and undecanal as Aldehyde C11 undecylic. Treatment of undecylenic acid with a strong acid causes the double bond to migrate along the chain by repeated protonation and deprotonation. Once a carbocation has formed on the fourth carbon of the chain, it can be trapped by the carboxylic acid group to form a stable γ-lactone, undecalactone, which has a very powerful coconut-like odour. Esterification of undecylenic acid followed by anti-Markovnikov addition of hydrogen bromide

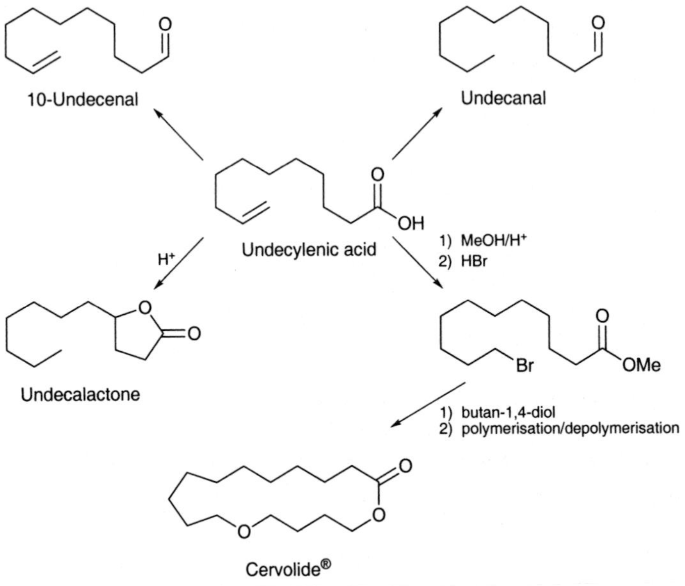

10-Undecenal

Undecanal

Undecylenic acid

H^+

1) MeOH/H^+
2) HBr

Undecalactone

1) butan-1,4-diol
2) polymerisation/depolymerisation

Cervolide®

Scheme 4.62

gives the ω-bromoester. Addition of butan-1,4-diol to this with subsequent polymerization and depolymerization (see page 99) gives the musk Cervolide®.

Synthetic precursors for aliphatic materials mirror the pattern of their naturally derived counterparts in that the commonest units are even in carbon chain length, because they are usually derived from ethylene through oligomerization. Thus, coupling of two ethylene molecules produces a four-carbon chain, three produces six and so on. To obtain an odd number of carbon atoms in the chain, one of the simplest techniques is to add a single carbon to an even chain; this can be achieved, for example, by hydroformylation. Hydroformylation also introduces an alcohol function and opens the way for oxidation to aldehydes and acids. Three carbon units are available from propylene as well as by reaction of ethylene with a 1 carbon unit. *cis*-Hex-3-enol occurs in a number of natural sources, such as freshly cut grass and strawberries. It possesses a very intense odour, which falls into the odour class known to perfumery as green. Green odours are those that resemble foliage, and stems of plants. *cis*-Hex-3-enol is very characteristic of cut grass and is used to add a fresh green topnote to fragrances. A number of its esters, such as the acetate and salicylate, are also of use.

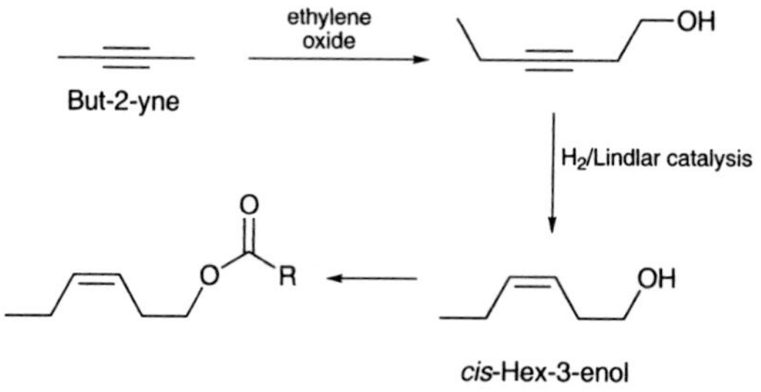

ethylene oxide

But-2-yne

H_2/Lindlar catalysis

cis-Hex-3-enol

Scheme 4.63

The synthesis of these materials is shown in Scheme 4.63. But-2-yne is obtained as a by-product stream from a petrochemical process. The hydrogen atoms adjacent to the acetylene bond are acidic enough to be removed by a very strong base, which allows the triple bond to migrate to the end of the chain. Once there, the terminal hydrogen atom is lost to give the relatively stable acetylide anion, and eventually all of the material is present in this form. The anion reacts with ethylene oxide to give *cis*-hex-3-ynol, which can be converted into *cis*-hex-3-enol by hydrogenation over a Lindlar catalyst. The same acetylide anion can also be produced from 1,2-butadiene which is another petrochemical by-product.

The Prins reaction, the acid-catalysed addition of an aldehyde or ketone to a double bond, is a useful reaction in perfumery chemistry and a number of aliphatic fragrance ingredients are prepared in this way. One example is shown in Scheme 4.64. When oct-l-ene is reacted with formaldehyde, the result is a complex mixture of products of which the pyran shown in Scheme 4.64 is the major component. In the case of this pyran, the octene molecule reacts with two molecules of formaldehyde to give an intermediate cation, which is trapped by the acetic acid present in the reaction medium. The product mixture is widely used for its fatty jasminic character and is sold under a variety of trade names, such as Jasmopyrane®.

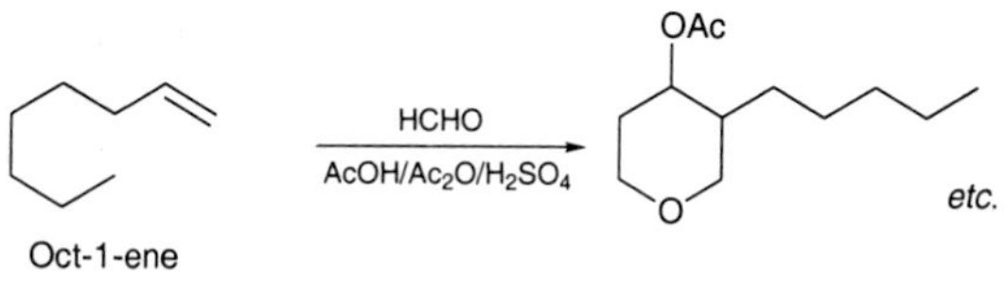

Oct-1-ene

HCHO

$AcOH/Ac_2O/H_2SO_4$

OAc

etc.

Scheme 4.64

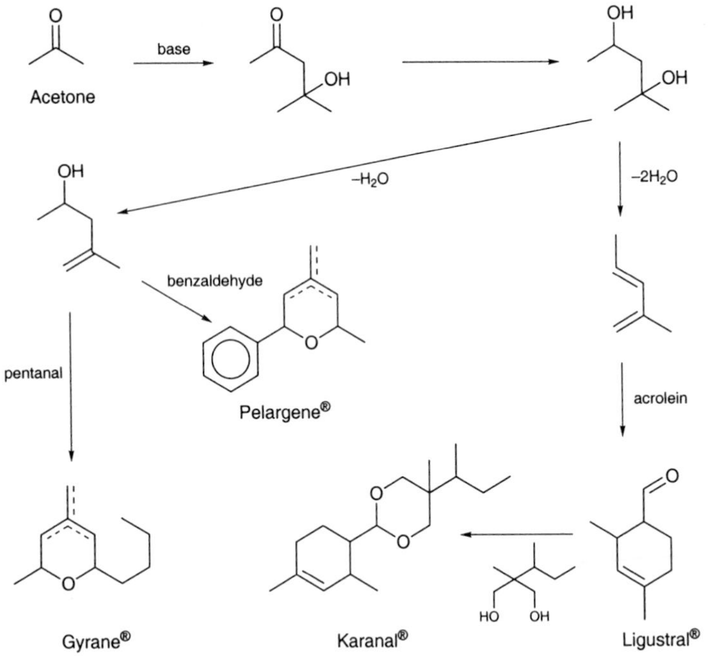

Acetone

base

OH

OH

OH

$-H_2O$

$-2H_2O$

benzaldehyde

pentanal

Pelargene®

acrolein

O

O

HO OH

Gyrane®

Karanal®

Ligustral®

Scheme 4.65

Two other Prins reactions are shown in Scheme 4.65. The precursor for both of these is 2-methylpent-l-en-4-ol, which is produced from acetone. Acetone readily undergoes a self-aldol reaction in the presence of base to give diacetone alcohol, 2-methylpentan-4-on-2-ol, which can be reduced to hexylene glycol, 2-methylpentan-2,4-diol. Careful dehydration under mild conditions gives the unsaturated Prins precursor. Dehydration under stronger conditions gives 2-methylpenta-1,3-diene, which is the more thermodynamically stable of the two possible diene products. Prins reaction of 2-methylpent-l-en-4-ol with pentanal gives Gyrane® and with benzaldehyde gives Pelargene®. In both cases, the cationic intermediate is trapped by the alcohol of the starting material to form a pyran ring. Both products have odours which are rosy and green: Gyrane® is fresh and radiant and Pelargene® resembles crushed leaves. Methylpentadiene is also a useful precursor for fragrance ingredients; it undergoes a Diels–Alder reaction to give Ligustral® (Scheme 4.65), a very intense green ingredient in its own right and also a precursor for Karanal®, a powerful ambergris material.

PERFUME INGREDIENTS DERIVED FROM CYCLOPENTANONE

The chemistry of synthetic jasmine materials was given an enormous boost in the 1930s when Nylon 66® was launched as a product. Nylon 66® is a polyamide prepared using adipoyl chloride and hexamethylenetetramine as monomers. The 66 in the name refers to the fact that there are 6 carbons in each type of unit that lies between the amide links in the polymer chain. Thus, adipic acid is the key feedstock for Nylon 66® and the introduction of the latter meant that the former became a basic chemical commodity. Pyrolysis of the calcium or barium salt of adipic acid produces cyclopentanone, and so the availability of large quantities of the acid meant that the ketone could also be prepared at low cost.

Cyclopentanone readily undergoes aldol condensation with a variety of aldehydes to give the 2-alkylidenecyclopentanones (Scheme 4.66). These have jasmine-like odours, but are no longer used in perfumery since it was discovered that they have the potential to cause skin sensitization. The saturated products are safe and are used to give jasminic, fruity, floral odours in fragrances. The most widely used are the *n*-heptyl- (R = pentyl) and *n*-hexyl- (R = butyl) derivatives. These are sold under tradenames such as Heptone® and Jasmatone®, respectively.

The 2-alkylidenecyclopentanones are readily isomerized to give the corresponding 2-alkylcyclopent-2-enones by the action of acids or platinum group metals. Michael addition of dimethyl malonate to these, followed by partial hydrolysis and decarboxylation, gives jasmonic acid analogues. The most important of these is methyl dihydrojasmonate, methyl 2-(2-pentylcyclopentan-3-on-l-yl)acetate. This material differs from methyl jasmonate, the natural jasmine component, in that its side chain is saturated, whereas the natural material has a *cis* double bond between the second and third carbon atoms of the chain. Both compounds have similar jasmine odours. The odour is perceived as weak when a fresh sample of material is smelled. However, if the sample is left in a room, the whole room is filled with its delicate floral scent. In perfume compositions, methyl dihydrojasmonate has a blending, fixing and enhancing effect on the other components. These properties have made it one of the most important fragrance ingredients.

When the 2-alkylcyclopentanones are subjected to the Bayer–Villiger reaction, the resultant lactones are found to have sweet, buttery, peach and coconut odours. These are useful in floral fragrances, but their major use is as butter flavour in margarines.

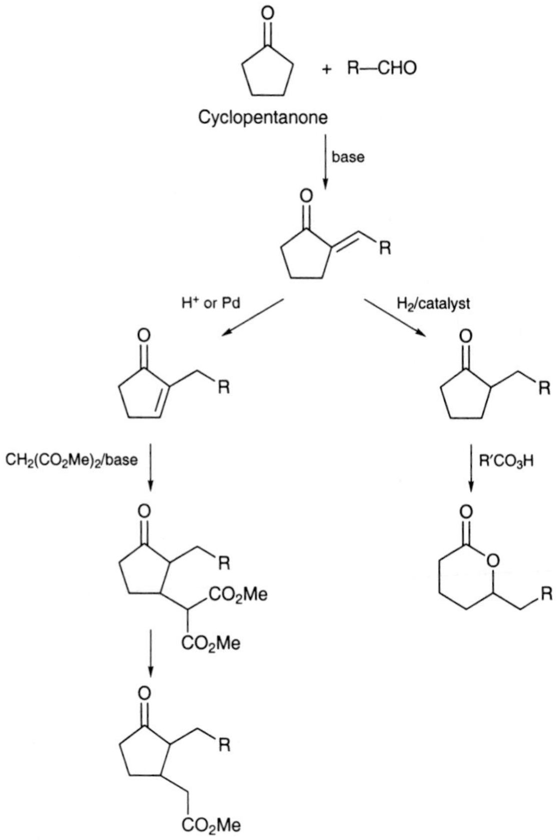

Cyclopentanone

+ R—CHO

base

H$^+$ or Pd

H$_2$/catalyst

$CH_2(CO_2Me)_2$/base

R′CO_3H

CO_2Me

CO_2Me

CO_2Me

Scheme 4.66

PERFUME INGREDIENTS DERIVED FROM DICYCLOPENTADIENE

Dicyclopentadiene is a feedstock for both the fragrance and polymer industries. It forms spontaneously from cyclopentadiene by a Diels–Alder reaction, and a *retro*-Diels–Alder reaction can be used to regenerate cyclopentadiene from it. A number of minor fragrance ingredients are produced by Diels–Alder reaction of the monomer with a variety of activated olefins in which the activating group X, is usually an aldehyde, ketone, ester or nitrile. However, the main fragrance uses stem from the dimer.

The two double bonds of dicyclopentadiene are very different in reactivity and, consequently, selective reactions are easy to achieve.

The bond in the bridged ring suffers a high degree of angle strain and readily undergoes any reaction that results in the two carbon atoms changing from sp^2 to sp^3 hybridization, since the latter have bond angles much closer to those required by the bridge structure. Thus, for example, carboxylic acids and alcohols will add to that bond under acid catalysis. The product is, of course, a mixture since the regiochemistry of the addition is unaffected by the position of the other double bond. The most important, by far, of the products shown in Scheme 4.67 is the product of addition of acetic acid, Jasmacyclene®. This material has a fruity, jasmine-like odour and its use by the fragrance industry runs into thousands of tonnes per annum. Apart from Florocyclene® and Gardocyclene®, the other products shown are all relatively low in tonnage. Dupical® and Scentenal® have fresh, floral, aldehydic odours, the others have varying proportions of floral and fruity notes.

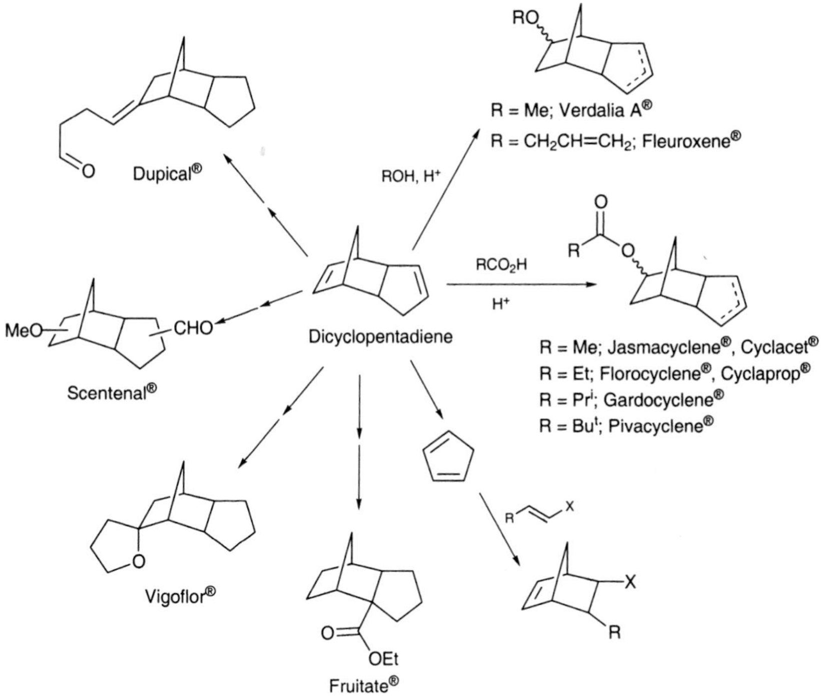

RO
R = Me; Verdalia A®
R = $CH_2CH{=}CH_2$; Fleuroxene®
Dupical®
ROH, H^+
RCO_2H
H^+
Dicyclopentadiene
MeO
CHO
Scentenal®
R = Me; Jasmacyclene®, Cyclacet®
R = Et; Florocyclene®, Cyclaprop®
R = Pr^i; Gardocyclene®
R = Bu^t; Pivacyclene®
Vigoflor®
OEt
Fruitate®
X
R

Scheme 4.67

CONCLUSIONS

The fragrance industry employs a very wide range of chemical processes and often has to deal with complex mixtures of structurally complex materials. The products must be produced at low cost and with minimal environmental impact. Minor components in mixtures may play disproportionately large parts in determining the odour of the whole, making quality control of processes a highly skilled occupation. The mechanism of perception of odorants is poorly understood in chemical terms. The continuously changing nature of the consumer goods' market demands a continual supply of novel ingredients to meet the challenges of new products to be perfumed. Initial bioassay (smelling) can be performed by the chemist and, thus, feedback is instant; also there is no need for lengthy activity testing as in the pharmaceutical and agrochemical industries. Performance testing and scale-up are also faster than is the case with those industries. All of these factors combine to make fragrance chemistry an exciting and challenging environment where scientific, artistic and commercial skills must all be practised together at the highest level of each.

REFERENCES

H. U. Daeniker, A. R. Hochstettler, K. Kaiser and G. C. Kitchens, *J. Org. Chem.*, 1972, **37**, 1.

B. A. McAndrew, S. E. Meakins, C. S. Sell and C. Brown, *J. Chem. Soc., Perkin Trans. I*, 1983, 1373.

P. J. Teisseire, *Chemistry of Fragrant Substances*, VCH, 1993.

G. Wagner, *J. Russ. Phys. Chem. Soc.*, 1899, **31**, 690.

H. Meerwein, *Ann.*, 1914, **405**, 129.

E. W. Spanagel and W. H. Carothers, *J. Am. Chem. Soc.*, 1935, **57**, 929.

L. Ruzicka and W. Brugger, *Helv. Chim. Acta*, 1926, **9**, 389.

M. Stoll and A. Rouve, *Helv. Chim. Acta*, 1934, **17**, 1283.

Chapter 5

The Structure of an International Fragrance Company

DAVID PYBUS

INTRODUCTION

Just as the senses of taste and smell, both chemical senses, are integrally connected, so the large, multinational fragrance suppliers are involved in flavour manufacture as well as that of perfume. Table 5.1 illustrates a near ten billion dollar industry in the provision of fragrance and flavour compositions, whilst Table 5.2 details the fifteen billion dollar market for cosmetics and fragrance on a global basis. Companies such as Quest are the ghostwriters of this industry. Their products, essentially fragrance and flavour concentrates are provided in 50 kilo, 100 kilo and 200 kilo drums, and occasionally in one tonne iso-containers. However, the multinational fragrance houses do not sell to the general

Table 5.1 *Estimated world consumption of flavour and fragrance products in 1994 (millions of dollars; adapted from SRI, 1995)*

	United States	*Western Europe*	*Japan*	*Rest of World*	*Total*	*Total (%)*
Fragrance composition	660	1190	250	720	2820	29.2
Essential oils/natural extracts	450	768	158	320	1716	17.5
Aroma chemicals	464	582	182	206	1434	14.8
(Flavour compositions)	820	1060	977	860	3717	38.5
Total	2394	3600	1567	2106	9687	100.00
Total (%)	24.7	37.2	16.4	21.7	100.0	

Table 5.2 *World consumption of cosmetics and perfumes in 1994 (adapted from SRI, 1995)*

	Billions of dollars	*Percent*
Cosmetics		
North America	7.4	35
Japan	6.0	29
Western Europe	4.5	22
Rest of the World and duty-free shops	2.9	14
Total	20.8	100.0
Perfumes		
Western Europe	7.1	48
North America	5.4	37
Japan and Rest of the World	1.4	10
Duty-free shops	0.8	5
Total	14.7	100.0

public. Their marketplace, through industrial marketing, is the large multinational companies that manufacture the well known branded products, or indeed, the multiple retail chains in any domestic economy that produce own brand products.

The six largest multinational fragrance houses account for some 70% of total fragrance compound sales. Theirs is a legacy of a century of experience and a financial strength that can cope with the tens of millions of dollars expense each year in pure research and development required to keep ahead of the game.

The structure and nature of a fragrance 'house' must reflect its link to the consumer *via* its own multinational clients. It is arguable that the client knows the marketplace best, and certainly invests in much research to ensure this is so, but the fragrance houses must understand sensory consumer perception and be alert to the reflection of fashion in odour terms. The marketeers in a fragrance house must be able to anticipate the effects of social trends on fragrance preferences and focus creative resources accordingly, relating brand values to odour and interpreting market research to understand precisely what is influencing consumer purchasing decisions. These companies understand the value of the psychology of scent.

Fragrance houses are fed a daily diet of perfumery briefs. Their survival depends on the successful nurture and assimilation of the brief leading to a winning response. They are thus structured around two main themes, namely:

—*The business-getting chain*. The proactive work on a client's brief with creative and technical teams developed to deliver winning products to the marketplace.
—*The supply chain*. The managed and planned purchasing of raw materials, competitively costed formulae, total quality production techniques and a customer delivery service department second to none.

THE BUSINESS-GETTING CHAIN

A fragrance house is typically structured in a three dimensional approach which covers geography, product type and functional skills. The core teams are structured around market product types (Figure 5.1), main groupings being Personal Wash (soaps and shower gels), Cosmetics and Toiletries (deodorants and shampoos), Fabrics and Detergents (laundry cleaners and fabric softeners), Household Products (multipurpose cleaners and air fresheners) and Fine Fragrances (essentially the high-profile alcoholic fragrances which so readily embody the general image of a fragrance house).

At the sharp end, marketing and sales personnel (account managers) relate with the client and help interpret the brief. This is fed into the wider team system for response within an agreed time-scale. A core team for brief response normally consists of people from Marketing, Perfumery (creative and technical) and Evaluation, whilst Research and Development may also be involved and technical advice given by the Product Applications Laboratory. As in all large companies, team working is the key.

THE SUPPLY CHAIN

The core areas of the client supply chain, involved with the formulation, material collection, production and supply of compound fragrances, are illustrated in Figure 5.2, with the key links between company departments shown. It should be emphasized that this diagram is not definitive. Each fragrance house displays a shape and structure that best suits its own marketplace, and there is a more complex interplay of communication than sketched here.

The scope and breadth of raw material requirement has already been mentioned in Chapter 3. Synthetic aroma chemicals, barring major oil crises, do not normally give problems in terms of stability of supply, relatively stable costs and reliable quality. Natural products are a different story, with the price and quality of an essential oil being very

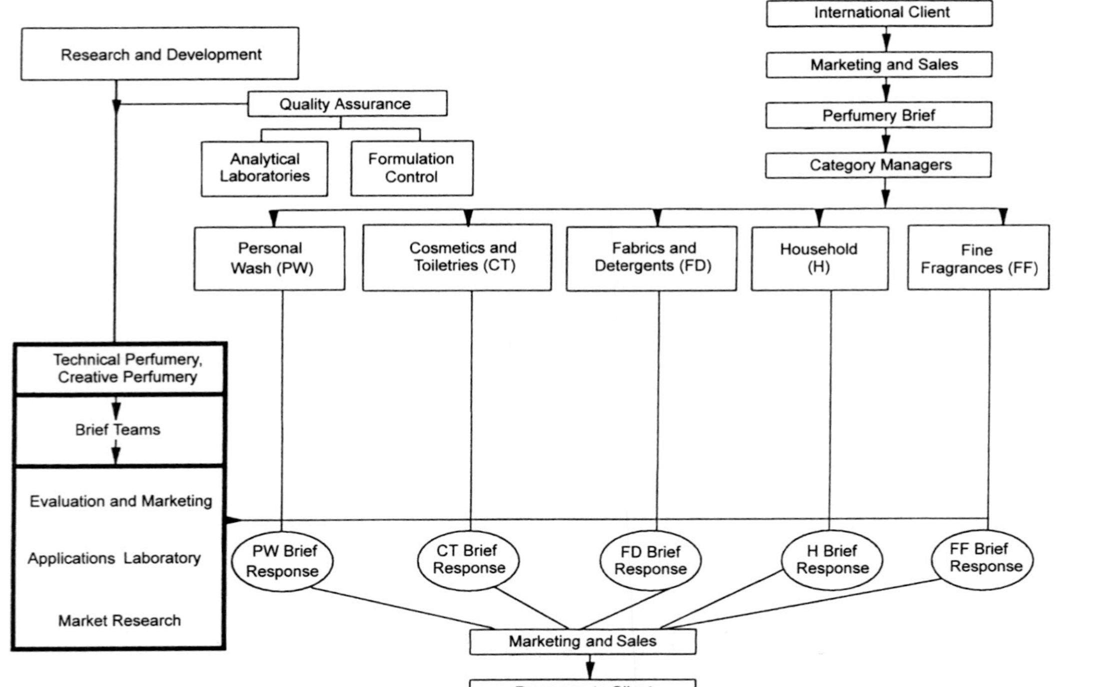

Figure 5.1 *General structure of an international fragrance company around the perfume brief (adapted from Curtis and Williams, 1995)*

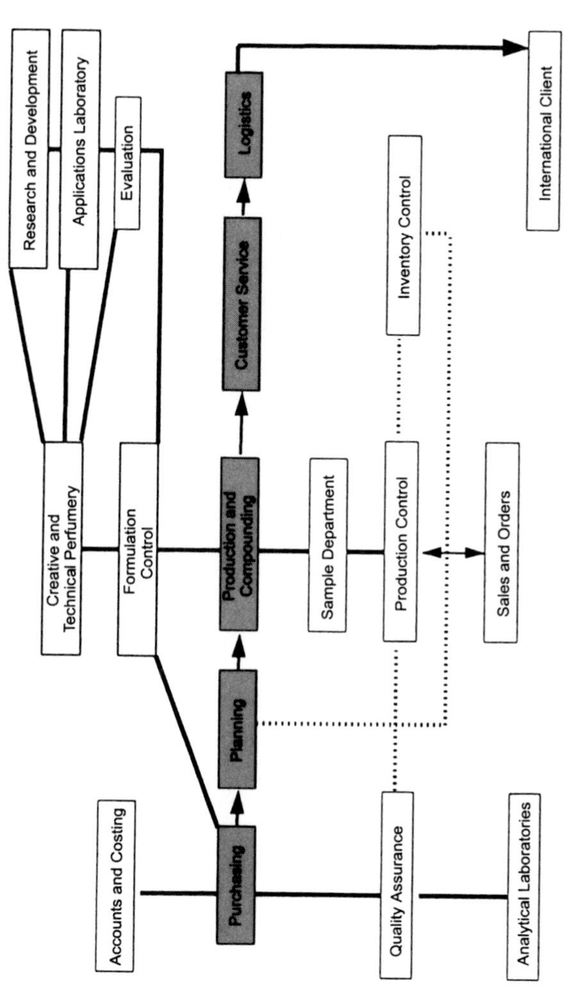

Figure 5.2 *General structure of an international fragrance company around the supply chain (adapted from Curtis and Williams, 1995)*

much related to 'acts of God', or the weather. Anyone who has some comprehension of the vagaries of a stock market centred on buying and selling futures in crops such as coffee and citrus fruits can multiply the concerns a thousand-fold to cope with the complexities of balancing stock, ensuring consistent quality and delivering finished perfume compounds on a competitive basis to the end client. The supply chain has two natural facets:

—Supply of standard orders on a regular, planned basis to clients.
—Supply of a newly-won perfume compound into the marketplace for the first time, and initial build-up of supply and demand balance set by the pull of market acceptance.

A fragrance house is like a bespoke tailor, supplying client-specific products which, for multinational clients, become essentially their own. There is, however, also an 'off-the-peg' service or 'fast-shelf' service, where the resource commitment, pricing demand and fast turnaround of response dictate a different method of interpretation.

REFERENCES

T. Curtis and D.G. Williams, *Introduction to Perfumery*, Ellis Horwood, 1995.
SRI, *Flavors and Fragrance Report*, SRI Consulting's Speciality Chemicals Update Programme, 1995.

Chapter 6

The Fragrance Brief

DAVID PYBUS

Chapter 5 illustrates how a fragrance house is structured around its lifeblood: the receipt, response to and winning of client briefs on a regular basis. Before developing a response to the Business Scents 'Eve' brief, it is worthwhile outlining what the main considerations of the brief are, as this reflects on the chemistry of the final fragrance choice and submission.

Briefs come in all shapes and sizes, and can be subject to many types of interpretation. Often the client is not exactly certain of what is wanted, or is not capable of translating the requirement into the language of perfumery. It is the work of a fragrance house to help the clients translate their gut-feel, or verbalize their inner sense of market opportunity into a viable fragrance direction for the product.

Thus, when a fragrance house is asked for the 'smell of an Arabian souk', it is well aware that the client means the fruit, spices and incense from such a scene, but naturally does not wish to include the inevitable environmental malodours which may come with the territory. A customer wanting 'the smell of the funfair' seeks chocolate, vanilla, fudge, toffee apples, candy floss, the cordite of the rifle range and the sharp ozone smell of dodgem car sparks.

The word 'fresh' can mean a multitude of different things depending on the language and culture in which we are dealing and, indeed, the olfactory experiences of the same.

In the main, however, the task is simplified by guiding clients by means of a printed 'Perfume Brief Document' or by catalysing their thoughts by using an *aide-mémoire*, which sales and marketing people carry in their faithful filofaxes! The majority of briefs received are not

Table 6.1 *Perfumer's rule of 13: an* aide-mémoire

The thirteen 'P's that constitute the building bricks of a substantial fragrance brief, with a general outline of the key thoughts linked to them, are:

1. **Product**	What is the product formula? What chemical environment will the perfume face? Does the product have colour, or any other physical characteristics such as base odour? Any special ingredients?
2. **Positioning**	What is the intended market positioning of the product?
3. **Place**	Is the product intended for the national, regional or global marketplace?
4. **Production**	What production process will the product undergo, at what time and how will it be dosed?
5. **Package**	Can the packaging (*e.g.*, aerosol canister lining, soap wrapper) be affected by the nature of the fragrance?
6. **Publicity**	On what platform is the product to be promoted? Are there key words, such as smooth, gentle, caring, hardworking, which the perfume will need to evoke?
7. **Pending**	What time-scale does the brief have? If very short, perhaps a shelf product will suffice; if long term and a significant brief, market research may be feasible and a prerequisite.
8. **Purpose**	Is the product new or a range extension? What competition, if any, is it up against? What is the key objective for the product?
9. **Price**	Cost per tonne of finished product is a better focus than cost per kilo of product, as it gives flexibility on dosage. Keep within the parameters set (relevant to fragrances for functional products).
10. **Presumption**	Ensure you understand the nuances of the country. What does 'green' or 'fresh' mean in the context of the customer's requirements? Never presume anything about any aspect.
11. **Pugilist**	Who are the competitive fragrance houses in the brief? What are their known strengths and weaknesses?
12. **Perfume**	Does your final submission meet all the criteria above? If the customer mentions a fine fragrance by name is that particular brand or fragrance direction really meant? Probe. How much fragrance is required for the brief testing? What regulations must the fragrance meet?
13. **Profit**	On submission, what margin (profit) suffices so that the winning of the business is worthwhile? A balance must be struck between what is realistically achievable and what is uncompetitive. Between what the client can afford and the true cost of what is being asked for.

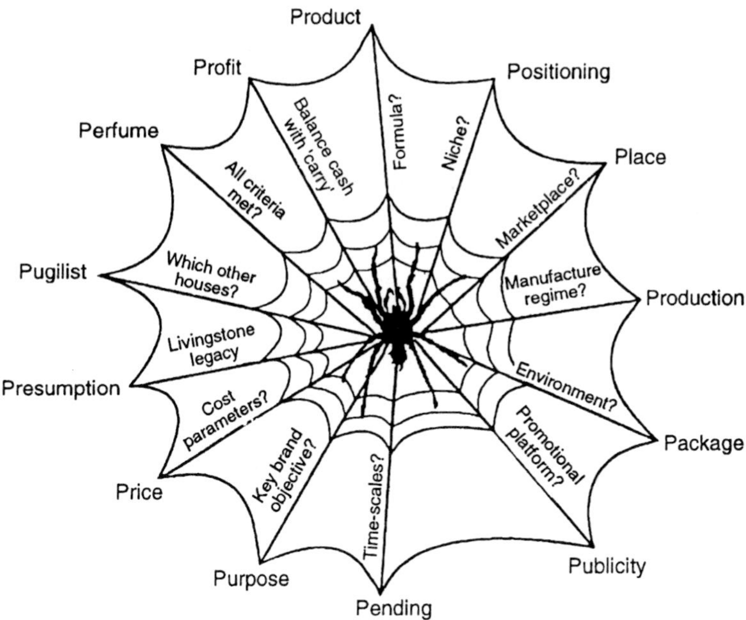

Figure 6.1 *The perfumer's creative web: 'tridecareglia' (the rule of 13)*

for exotic fine fragrances, which are low-volume, high-value perfumes, but for the low-value, high-volume perfumes required for functional products (such as personal wash or household brands). Figure 6.1 and Table 6.1 give a brief description of the kinds of response and background that are being sought, in both list and visual form. The idea is that the more is known, the less likelihood of error and the more likely a successful win in the brief.

These initial focii for thoughts are, of course, only the beginning of a process that to be truly successful involves continual and frank discussions with a client on an ongoing basis.

It is along the lines developed in Figure 6.1 and Table 6.1 that following early discussions with Business Scents Ltd, the brief outlined below was received.

BRIEF FOR 'EVE', PREPARED BY BUSINESS SCENTS LTD

Background

Business Scents Ltd wish to launch a new female fine fragrance this year that will complement the successful launch of the *Ninevah* range, launched in 1994, and essentially the hope is that a roll out will occur in Europe and the USA, followed by Asia Pacific, over a two-year time span. If successful, it is intended to add line extensions to the range, code-named 'Eve', to include an antiperspirant, soap, shampoo and shower gel. Its global market requires any fragrance submitted to have 'global appeal' to the core market of females aged between 16 and 35 years, and we look to the fragrance house to suggest also possible names and marketing platforms.

Product Range

1. *Alcoholic fine fragrance*. Contains 90% denatured, cosmetic-grade ethanol. Possibly lightly coloured (maybe a pastel shade).
2. *Soap*. A predominantly vegetable base (80% palm, 20% coconut). Likely to be white, but could be coloured to support the fragrance concept.
3. *Antiperspirant*. Aerosol form using butane–propane blend as propellant, and packed in lacquer-lined aluminium cans. Should contain an active antiperspirant at an effective dosage (*e.g.* aluminium chlorhydrate).
4. *Shampoo*. Mild detergent system with built-in conditioners. Suitable for frequent use. Will likely to be pearlized and coloured to support the fragrance. Packaging: polyethylene terephthalate (PET) bottle.
5. *Shower–bath gel*. Clear, high foaming and capable of moisturizing the skin. Any advice on novel and useful botanical extracts? Packaging: low density polyethylene or PET.

General

Product formulation advice is sought from the fragrance houses for products 2–5 to aid with initial stability-testing work and to guide the contract packer(s) with their development.

Fragrance

We seek a novel approach to the fragrance for 'Eve'; perhaps even with new aroma chemicals, or at least an intriguing marketing platform that greets the new age. While *Ninevah*, our 1994 launch, has been extremely successful we are averse to a fragrance that is in the same fragrance area (rose–jasmin, floral, woody, moss, oriental) as the brand code-named 'Eve' is expected to strike a new direction and signal the new challenges of the Millennium. Along with a fragrance concept we are open to thoughts on both name and a broad marketing platform for the perfume.

Time-scale

A submission is required (maximum one kilo) within six months of the brief date. Meantime, constant contact with Marketing at Business Scents is encouraged to determine whether the directions being reviewed are realistic. Interim submissions and dialogue are suggested. We require only one submission per fragrance house, although it is likely that we will develop both the theme and direction of the fragrance at a later stage with the house chosen from this initial brief.

Brief Recipients

There are four recipients for this brief, namely Quest International, Smells 'R' Us, Panaroma and OL-FACTORY.

Price

We do not envisage a cost in excess of US$50 per kilo on a delivered basis to our main manufacturing unit in Holland for the fine fragrance perfume. Modified perfume in the same fragrance direction, but workable in appropriate bases, should cost no more than US$30 per kilo for soap, shampoo, shower gel and antiperspirant formulations.

Production

You are familiar with all our production methods for fine fragrances, soaps, shower gels, shampoos and antiperspirants, having had the details of equipment from the *Ninevah* brief of 1994. Direct contact with our contract packers is recommended to review and update your knowledge of these methods, and how they impact on perfume dosage.

We look forward to dialogue with your company over the next three months, and to your final submission to this exciting, 'end of century' project.

Chapter 7

Perfume Creation: The Role of the Perfumer

LES SMALL

Business Scents Ltd has briefed Quest to repeat their success of the 1990s, in which a popular fine fragrance launch was followed by a successful launch of a series of trickle-down products, namely a soap, a shampoo, a shower gel and an antiperspirant. The brief is known as 'Project Eve'.

Market research indicates that an exotic, tropical, fruity, watery, floral, muguet (lily of the valley) theme would be a likely winner (see Chapter 8).

PERFUMERY DEGREES OF FREEDOM

At the outset a perfumer starts with a blank sheet of paper (or an empty computer screen). From where come ideas and inspiration? What is currently fashionable? The name of the game is to imagine the odour effect that the Business Scents product will exhibit in use. How should the alcoholic product smell from the bottle or atomizer? How will it develop on the skin once applied? How long should it last between applications? Once these decisions have been made the perfumer can start compiling a list of the fragrance raw materials that may be used to achieve the desired effect. At first sight, there appear to be hundreds of ingredients to choose from, but in practice there are a number of restricting factors. This concept of freedom of choice of ingredients can be described by employing the analogy of Degrees of Freedom as used in physical chemistry in which a system is constrained by a number of

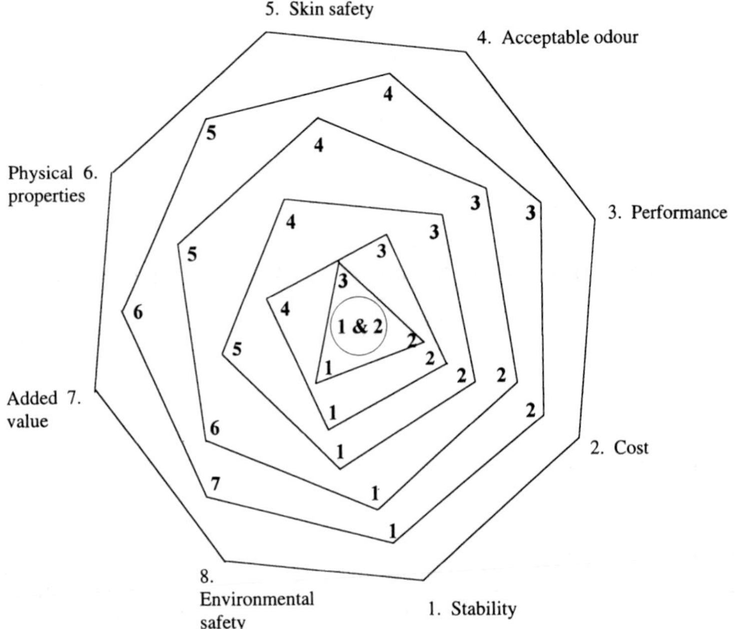

Figure 7.1 *Perfumery degrees of freedom*

factors. The EIGHT degrees of freedom in perfumery are shown in Figure 7.1 and listed below:

—SKIN SAFETY.
—ENVIRONMENTAL SAFETY.
—ACCEPTABLE ODOUR.
—COST.
—STABILITY.
—PERFORMANCE.
—PHYSICAL PROPERTIES (*e.g.* solubility, colour).
—ADDED VALUE (*e.g.* malodour counteractancy, insect repellency).

Figure 7.1 is a series of concentric polygons; it is an attempt to visualize the concept of the perfumery degrees of freedom. The outer octagon represents the system with no constraints, *i.e.* all eight degrees of freedom are available. If a constraint is introduced, for instance *stability*, then the polygon becomes a heptagon and the area contracts. This is analogous to restricting the number of ingredients that can be employed in the creation of the masterpiece! As further constraints are introduced one by one, the polygons become smaller and the number of sides diminishes accordingly. Eventually, one arrives at a small area in the centre where nearly all of the degrees of freedom have been lost. This is the situation in which relatively few perfume ingredients are available owing to the hostile nature of the medium to be fragranced, for example a limescale remover based on sulfamic acid. Perfume can often provide *added value* to a product mix. For example, it is possible to provide deodorant protection for a soap or deodorant–antiperspirant by combining certain chemical classes of ingredients, according to a patented set of rules. (See Patent Filings 1980/1981.)

THE ALCOHOLIC FRAGRANCE

An alcoholic fragrance provides almost complete freedom of choice; there is the possibility of acetal formation due to the interaction of the aldehyde and ketone functional groups, with the alcohol used as the carrier, but this often has a softening effect on the fragrance and is not perceived as a disadvantage. No other serious chemical constraints are placed on the choice of ingredients, *i.e.* most of the ingredients are *stable* in the aqueous alcoholic product. As discussed later, stability plays a major role in the choice of ingredients for an antiperspirant product.

The *safety* constraint has been defined in the brief and can be checked by computer analysis or by the safety officer at any time.

One of the ways in which ingredients can be classified is by their perceived odour properties. The most important facet, as far as the perfumer is concerned, is the odour description. Each perfume company tends to have its own system of classifying odour, but in general there are invariably a set of odour families that tend to be subdivided into odour descriptors. Other important properties are odour intensity (and how sensitive the intensity is to dilution) and longevity of the ingredient on a relevant substrate; in this instance, the skin. During the development phase the relative longevity of initial perfume trials is compared on smelling blotters (thin, absorbent paper strips); only towards the end does the perfumer fine tune with trials on the skin. This is for practical reasons such as keeping the immediate environment as odour free as possible, and that space on the perfumer's arms is limited!

A search is made for ingredients that have the desired odour characteristics. For the Business Scents Ltd brief, market research indicates that exotic, tropical, fruity and watery floral muguet notes are desirable. This is the *acceptable odour* (hedonic) constraint in the degrees of freedom analogy. An initial trawl through the various databases indicates several-dozen contenders. Many of the tropical, fruity notes will have an ester or sulfurous functional grouping, whilst several aldehyde groups exhibit the desired watery muguet character, *e.g.* Lilial®, Lyral®, cyclamen aldehyde, Bourgeonal® and hydroxycitronellal. Interestingly, a perfumer can often distinguish the typical odour of certain functional groups. Salicylates tend to have a woody character, which is modified by the length of the attached aliphatic chain or ring structure. Nitriles often smell somewhat metallic. Phenyl acetates usually exhibit aspects of honey, and so on.

The perfumer usually consults colleagues in the Natural Products Analysis department to establish new ideas for natural tropical, fruity notes (see Chapter 12). They provide some headspace analyses that should add interesting, novel top-notes to the fragrance creation … in this case a headspace analysis based on a selection of tropical fruits purchased from a market stall in Jakarta, Indonesia. Having checked with Quest's Buying Department as to the commercial availability of the chemicals reported in the headspace analysis, it was found that several of those available had already been incorporated into a perfume called *Tropical Cocktail.* This has been used as a building block at approximately 2% in the final fragrance mix. Details of the headspace analysis of muguet and broom were also provided and the decision made to use a small quantity of broom absolute in the 'Project Eve' fragrance, to add a natural, fresh florality which would fit well with the watery muguet theme.

So, having collected a set of watery, floral ingredients and some exotic, fruity ingredient ideas, the perfumer starts to place the ingredients together in groups. The watery ingredients could be blended with other light, floral and woody ones. Similarly, an accord of exotic fruity materials may be combined with other suitable odorous ingredients, such as musk and citrus notes, for example bergamot and grapefruit. These two accords are mixed in various ratios until a satisfactory blend is achieved. Other ingredients can be added one by one to this composition to modify the effect. For example, touches of camomile oil add an interesting twist to the fruity top-notes, whilst a small quantity of Amberlyn® provides subtle amber end-notes. The process is iterative in that a trial formula is made, assessed on a smelling blotter, the formula adjusted, reassessed, and so on. At all times the overall evaporation profile of the fragrance is considered, together with any *cost* and *skin safety* constraints set by the customer. The ingredients used have a range of volatilities and intensities, and need to be blended carefully together so that no single ingredient dominates the effect of the others. The overall effect, as the perfume evaporates on warm skin, should be a gradual change from the volatile top-notes of the fragrance through the middle-note theme to the eventual end-notes retained on the skin. The 'shape' of the perfume should be maintained at all times; so, for example, as soon as a volatile note such as a fruity ester has evaporated there is a less volatile fruity note available to maintain the fruity theme in the middle period of the dry-down. Similarly, at the end of the fragrance life-cycle on the skin, still less volatile fruity notes (*e.g.* lactones) should be available. In the same way, the green muguet floral notes should keep their shape throughout the evaporation period. For an alcoholic fragrance, the top-notes approximately represent the first 15 minutes of the evaporation; the middle-notes account for the next 3–4 hours and constitute the heart of the fragrance, whilst the end-notes represent the final 5–8 hours and give tenacity and depth to the theme.

DEVELOPMENT OF OTHER PRODUCTS

Once the fragrance has been evaluated on skin and approved by the evaluation panel, the perfumer can start to make versions for the other products in the Business Scents range; namely, soap, antiperspirant, shampoo, and shower and bath gel.

The first problem to be encountered is almost certainly one of cost. The client will not be willing to pay the same amount of money for the perfume oil used in the range as for that used in the alcoholic fragrance.

Additionally, each of these products has associated with them different challenges to the perfumer, who will seek the advice of many colleagues within the company. Some of the problems with each product are discussed next.

Soap

If the product is to be a white soap, will any of the ingredients in the perfume formula cause discoloration, either immediately or in time? How will the total perfume perform in the soap bar? Will it cover the fatty smell of the base? If not, which ingredients perform best in the dry bar and in use? To overcome some of these problems, certain ingredients in the alcoholic fragrance need to be substituted. For example, to prevent discoloration on storage, vanillin may be substituted by ethyl vanillin which, because it is more intense, can be dosed at about one-third the vanillin level. Small quantities of Ultravanil® are also incorporated to boost the vanilla effect without causing discoloration.

Indole is replaced by Indolal® at a slightly higher level, as it gives a less intense, animalic jasmine character in soap. Other ingredients need to be substituted because of their relatively high cost contribution to the perfume formula.

However, since many natural oils and absolutes also possess a correspondingly high odour strength and quality, the ratio of the cost to strength and quality should always be considered carefully before removing them. In this case, it is difficult to justify inclusion of the broom absolute in the soap context, so in its place a small quantity of methyl anthranilate (the chemical that quantitatively dominated the headspace of broom) is added. Care should always be taken with anthranilates because of the possible formation of Schiff's bases upon reaction with aldehydes. Extra aldehydes will certainly be added to the soap version of the alcoholic perfume because of their excellent odour performance in covering the fatty smell of common soap bases.

Shampoo

Discoloration and performance issues are important. For example, will the perfume diffuse from the shampoo pack and then again as the hair is being washed in warm water? Will it be substantive on the dry hair? Will it be soluble in the shampoo base? In Chapter 11, the issues relevant to predicting the performance of fragrance chemicals in certain situations are discussed. Molecules with a low relative molecular mass (RMM), for example ethyl hexanoate and limonene, diffuse most speedily on opening the cap of the shampoo bottle. Furthermore, depending upon their log P value, they are more or less 'happy' in the mainly aqueous environment of the shampoo: the higher the log P value, the more hydrophobic the molecules and *vice versa*. Hence, molecules such as limonene with relatively high log P (4.46) diffuse more readily than more polar molecules of similar RMM, such as 2-phenylethanol (phenyl ethyl alcohol or PEA) whose log P is 1.52.

Other factors complicate the situation and must be taken into consideration. An important one is the odour intensity of the molecule. A few very potent molecules in the headspace can give a greater odour appreciation than a large number of low intensity molecules. Another consideration is the active detergent level of the shampoo, which affects both the appreciation of the fragrance in the headspace above the shampoo and the solubility of the perfume oil in the detergent system. A system with a low concentration of active detergent has a smaller reservoir of micelles in which to solubilize the perfume mixture. Thus, more perfume is available for the headspace because less can be 'dissolved' in the shampoo base.

Substantivity on the hair is another important consideration when creating shampoo fragrances. The consumer expects his or her hair to be perceived as clean and fresh. This impression is closely linked to the longevity of the perfume ingredients on the hair. In this case, the molecules with high RMM values tend to be the ones that remain on the hair after rinsing and drying. Again, the odour intensity must be sufficiently high for detection by the nose as the number of molecules laid down on the hair surface is relatively small. Chapter 12 gives clues as to which fragrance molecules are most likely to be retained by hair protein, rather than disappear down the plughole along with the foam! Low $\log P$ value molecules tend to stay with the water. In fact, the perfumer also relies on substantivity data obtained by painstaking, empirical, ingredient studies. Dozens of hair switches are washed, rinsed, and dried following a protocol developed to distinguish between the good, poor or indifferent ingredients when assessed by a panel of perfumers. This information is then added to the perfumer's knowledge base for future use.

So, for the shampoo version, the same changes to avoid discoloration are made as for the soap version. However, the percentage weight of citrus oils is increased to boost the limonene content. The fruity notes are also made more dominant, as these are favoured by many shampoo consumers. It is necessary to check that the watery, floral and tropical fruity notes are detectable on the hair after the shampooing process.

Shower and Bath Gel

Again, issues of discoloration and performance are important. In addition to hair substantivity, retentivity on the body must be considered. In general, the perfume dosage in a shower gel or bath product is double that for shampoo, so the effect of any potentially discolouring ingredients is accentuated. The higher perfume level certainly helps fragrance retentivity. However, substantivity on skin is usually less than that on hair for three reasons. Firstly, the skin is warm and hence the rate of perfume vaporization is higher. Secondly, the skin continually secretes other chemicals onto the surface, which can interact with deposited fragrance. Lastly, the skin is less porous in nature than hair, so potentially fewer binding sites are available for the perfume molecules. If the product is destined for the bath, then diffusion from the bath water becomes important.

Antiperspirant

Here, the problems are associated with possible discoloration, and stability with the acidic aluminium chlorhydrate (the active ingredient). In Chapter 15, the reasons for the poor stability of Lilial®, hydroxycitronellal and cyclamen aldehyde in such a medium are explained. All three are present in the 'Project Eve' fragrance. In Chapter 10, the empirical stability testing conducted on a range of perfume ingredients in antiperspirant media is described. This knowledge base is available for screening the 'Project Eve' perfume for possible problem ingredients. Bearing this in mind, it is necessary to substitute the Lilial®, hydroxycitronellal and cyclamen aldehyde in the perfume formula with Florosa®, together with a small quantity of Bourgeonal®. Florosa® is very stable in antiperspirant and, although Bourgeonal® is not totally stable, it is so intense that only traces are required to deliver the desired watery floral effect. Similarly, other relatively unstable ingredients are substituted by more stable, better performing materials, whilst the overall theme or shape of the composition is always maintained.

CONCLUSION

So, it can be seen that the role of the perfumer is a blend of artist and scientist, continuously attempting to bring an imagined odour into being via a written perfume formulation in the same way that a composer imagines a symphony and realizes it by writing the notes as a score. As in all creative processes, a period of reflection is required prior to starting work. To create a pleasant fragrance in a reactive medium, the perfumer must have a full understanding of the odour, and physical and chemical properties of the available raw materials.

REFERENCE

Deo Patent Filings. Soap Bar/Deoperfume: GB2016507/US4288341, 1980/1981.

Chapter 8

Measurement of Fragrance Perception

ANNE RICHARDSON

INTRODUCTION

A wide range of factors affect how we perceive a product. If we are to understand the reasons behind why the consumer chooses a product, we need to determine what drives acceptance and how this can be interpreted into information that can be used to develop winning fragrances.

Sensory analysis is concerned with quantifying human responses to stimuli. It is a precise, descriptive and measuring technique that characterizes the stimulus. In this case, the particular concern is to evaluate the odour of a perfume, perfume ingredient or perfumed product. This is an important process in enabling the perfumer to understand and quantify the sensory characteristics of the product, as only then can they be manipulated in a controlled way as part of the creative process.

The evaluative and subjective associations made by the consumer must be understood when assessing a product; these are measured using market research techniques. If the market is understood, fragrances can be developed to match or enhance the image of a particular product or market segment. Sensory analysis is also an important tool in this process. Using powerful statistical techniques, the odour relationships between different products or perfumes can be characterized and quantified, and the results combined with market research to enable the subjective associations to be interpreted in odour terms.

Sensory analysis and market research rely on verbal or conscious measurement of an odour or perfumed product by a human respondent. Emotion is another aspect of product perception which is difficult to measure using these verbal techniques, yet can also strongly influence how a consumer perceives a product. Odours evoke emotional responses, which may take the form of bringing back memories of situations past or simply evoking feelings such as warmth or comfort. Research is currently focused on understanding how the brain interprets the olfactory messages to convey these emotions, and ways of measuring these responses.

In answering a customer brief such as that from Business Scents Ltd, a perfumer welcomes as much guidance as possible in how to win that brief successfully. The four disciplines mentioned above (sensory analysis, market research, statistics and psychology) together form a powerful analytical and predictive tool, different aspects of which can be used for guidance in perfume creation depending on the requirements of the brief. In the sections that follow in this chapter, a brief outline of the techniques currently used within each of these areas of expertise is given, as are examples of how they could be used to fulfil the Business Scents brief.

MARKET RESEARCH

Although consumers are able to state which perfumes or products they prefer, they are often unable to explain why. The skill in good market research lies in the design of questions that are able to find out as much as possible about the reasons behind the consumer preferences by asking how the fragrance relates to situations within the consumer's everyday experiences. For instance, if the product is a bar of soap, we could ask whether the respondent pictured the person most likely to use the soap as a young teenager or a middle-aged housewife, or we could ask whether the soap would be appropriate to use in the morning, before going out in the evening or before going to bed at night. Questions of this nature can be answered by the consumer and related to the product preference, which helps us to understand the important factors determining the preferences.

Unfortunately, consumer research is expensive. A good test requires a large population of respondents, and there is a limit to the number of samples each can test; typically, a test which includes 20 products or perfumes may require interviews to be carried out with 1000 respondents, each testing four products (giving 200 assessments of each product). The survey respondents must represent the target population and, as far as possible, the survey must be carried out under realistic circumstances. For instance, a laundry powder perfume would be tested by a population who normally do the household shopping and washing.

The perfumes would not only be smelled in the powder, but also on cloth that had been washed in the powders, and may even be taken home by the respondent and used as part of the normal washing routine.

To broaden the scope of market research tests, several additional types of test have been developed for use alongside the traditional, large-scale tests.

Focus Groups

Focus groups provide in-depth qualitative information about the product in test. They are led by group leaders trained to direct discussions within small groups of respondents as they smell the products. Different techniques can be used to explore associations that arise while smelling the products; for instance, by matching the products to different pictures or colours or by asking the respondents to describe types of people that they would associate with the smell. These groups are often also used without the stimulus of any products present, to develop totally new ideas for a product range.

Qualitative Descriptive Analysis

Qualitative Descriptive Analysis (QDA) is a technique developed using trained, in-house panels of assessors. The panellists are presented with the range of products to be assessed and spend their first panel session discussing the properties that they feel are most important in describing the products and the differences between them. These usually include evaluative attributes, such as refreshing or old fashioned, as well as odour descriptions, such as flowery or medicinal. The panellists agree a list of attributes and may also agree standards to help them describe the attributes. A series of panel tests is carried out, during which each panellist scores the products on each attribute in the agreed list. The results are averaged across the panel to give a QDA profile of each product, which may be presented as charts or graphs, or analysed in more detail using multivariate techniques (described later).

SENSORY ANALYSIS

Sensory analysis involves using human subjects as a measuring tool. This presents an immediate problem, as individuals are innately variable, not only as a result of their experiences or expectations, but also as a result of their sensitivity. Thus, each person could genuinely perceive the same product quite differently. It is therefore essential in every sensory test that all variables except that actually under test are as carefully controlled as possible to minimize this variability.

A purpose-built sensory panel suite enables testing to be carried out in an environment free of noise or movement, and in rooms that are painted a neutral colour and are fitted with individual booths for each panellist. When testing is carried out, all samples are presented unlabelled so that the panellists do not receive any cues concerning the nature of the product, and the sample appearance is identical; the only variable (as far as possible) is the smell.

Having controlled the environment and the sample, the only other major variable is the panellists themselves. To minimize panellist variability, all panellists are selected for their sensitivity to smell and their short-term odour memory. They undergo six months training to be able to recognize (and name) a range of odour standards, to score perceived intensity of odour and to use a range of standard sensory tests. These are all highly specialized skills which the average person in the street simply does not possess.

Odour can be described using a number of different dimensions, each of which can be measured using different sensory tests: threshold, intensity and quality.

Threshold

Threshold itself can be described at three different levels:

—Detection threshold. Can an odour be detected?
—Recognition threshold. Can an odour be identified?
—Difference threshold. Is there a difference between two odours?

The detection threshold is the lowest stimulus intensity (odour concentration) that the subject can distinguish from an odour-free situation. The subject's response indicates whether the presence of an odour has been perceived or not. Correspondingly, the recognition threshold is the minimum concentration at which an odour can be identified.

Such data provide important fundamental information, such as identifying levels at which an ingredient can be perceived in a product base, or quantitative information describing the activity of odour materials.

Intensity

How strong is an odour? A number of different types of sensory scales are used to measure perceived intensity of odour, two examples of which are given in Figure 8.1. Alternatively, panellists can be trained to use a form of scaling whereby they alot their own scores to the perceived intensity of odours and score subsequent odours in ratio to one another; this is known as ratio scaling, and it allows a more flexible and arguably more accurate measurement of perceived intensity.

When measuring odour intensity it is important to be aware of Stevens's law, which states that equal changes in stimulus magnitude (S) produce the corresponding change in perceived intensity (I). The law can be expressed as a power function, as in equations (1) and (2);

$$I = cSk \tag{1}$$

or

$$\log I = k \log S + \log c \tag{2}$$

i.e. the log of perceived intensity is directly related to the log of the stimulus magnitude.

The increase in perceived intensity with concentration can be represented by a straight line, as shown in Figure 8.2, for odorants A and B. The slope indicates how fast odour intensity rises with concentration and the intercept defines the detection threshold. There are two important characteristics of this type of data:

—At higher perfume concentrations a larger increase in concentration is necessary to give the same change in perceived intensity.

—At different concentrations of two odorants the rank order of their perceived odour intensity can change. This is true for odorants A and B in Figure 8.2. At log Cy, odorant A is stronger than odorant B but at log Cx odorant B is perceived as stronger than odorant A.

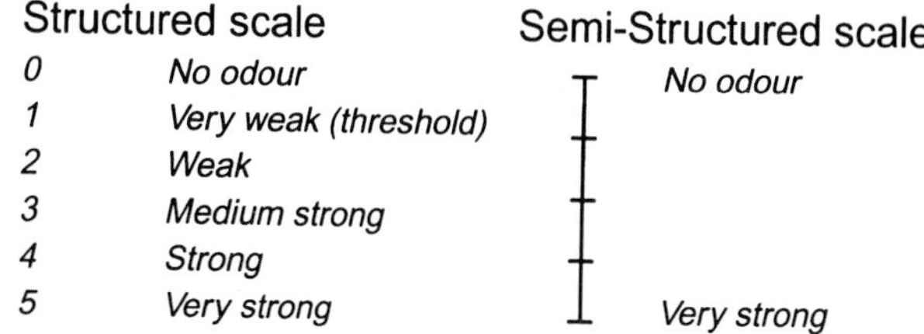

Figure 8.1 *Intensity scales*

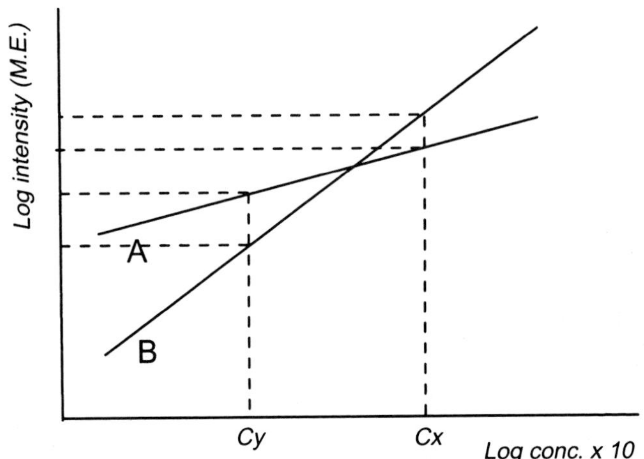

Figure 8.2 *Log perceived intensity* vs. *log perfume concentration for two odorants*

Quality

What does this odour smell like? The tests used to describe odour quality are known as odour profiling tests. These are the most complex of the sensory tests and, to ensure good quality, accurate and reproducible data are only carried out by highly trained and experienced sensory panellists.

Each odour is smelled by the panellist, who then scores the perceived intensity of each odour character that she or he can detect (referring to the set of standard odour references for clarification if necessary), which results in a sensory profile for that odour. A minimum of at least 20 profiles is usually collected for each sample and an average profile is then calculated. A set of typical odour profiles is shown in Figure 8.3. These profiles show the differences in perceived intensity of 13 odour characteristics identified in seven perfume materials, and immediately it is possible to see that although all of the materials are floral or muguet in character, one material is far more fruity (cyclamen aldehyde) and another (Mayol®) is far more herbal than the other materials.

Sensory profiling techniques are designed to produce stable and reproducible data, but difficulties arise when trying to compare data obtained from different laboratories. Often the methods of sensory assessment differ and there is no universally accepted odour language or list of odour standards to clarify this problem.

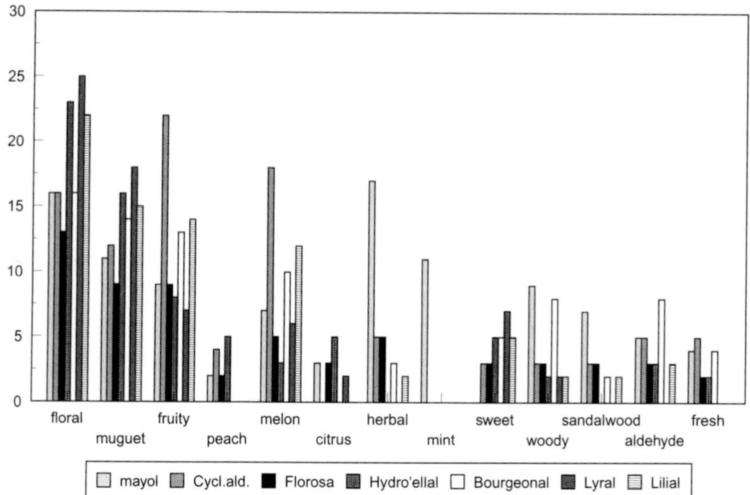

Figure 8.3 *Odour profiles of a selection of muguet perfumery materials*

Statistical Techniques

Fragrance profiles are often presented as a series of complex tables or graphs from which patterns, order or exceptions need to be found. Looking at individual profiles it is possible to determine the main odour characteristics. Comparison with the other profiles establishes the main differences between them.

However, this sort of comparison becomes extremely complex when large numbers of perfumes are involved. Multivariate analysis methods are descriptive procedures that help in this process. These methods are used to model or describe data such that they can be more easily understood by the researcher and thus simplify data comparison. Sophisticated software now exists that makes this kind of analysis possible without it being necessary to understand fully the mathematical modelling involved in the analysis. However, to interpret and understand the results a basic understanding of the technique is necessary.

Multidimensional Scaling

Multidimensional scaling is just one of the multivariate techniques available. To be asked to take a map of England and measure with a ruler the distance between 20 towns is a fairly straightforward project. Multidimensional scaling does the opposite to this; it takes a set of distances and recreates the map. The 'distances' in this case are derived from our sensory panel data by comparing the profiles of every possible pair of samples and deriving a value which represents the overall similarity between each pair. The resulting map places the samples spatially, so that those that are most similar to each other are shown closest together on the map; as samples become progressively more different they are shown progressively further apart.

Principal Component Analysis

Principal component analysis (PCA) is another technique used; it works differently, but the resulting map can often look very similar. The analysis works using a process of identifying correlations between the different variables used in describing the data (*i.e.* the sensory descriptors used in the profiles). It then searches for the combination of variables that best describes the maximum amount of variation in the data, and draws an axis through the centre of the group of observations, so that the sum of the residual distances is minimized; this is called the first principal component. The second principal component is then drawn at right angles to the first and explains the maximum amount of the remaining variation.

In addition to depicting the associations among the original variables, PCA can be used to describe the relative 'locations' of the measured samples. A plot of principal component scores for a set of products reveals groupings of the samples that may not have been readily apparent from the original data.

Figure 8.4 is a multidimensional scaling map derived from the profiling data shown in Figure 8.3. With only seven samples to compare, this is a very simple map; however, it serves to illustrate the kind of information that can be derived from this type of analysis.

The samples are grouped so that those that are most similar in odour character are closest together on the map (*e.g.* Lilial® and Bourgeonal®), and those that are most different are furthest apart (*e.g.* cyclamen aldehyde and Lyral®). The arrows indicate the direction of increasing perception of the odour characteristics shown. Only the odour characteristics that significantly correlate with the distribution of the samples across the map are shown; these are the characteristics that are responsible for the systematic differences between the samples.

All the materials were perceived to be floral and muguet in character, so these characteristics are not shown on the map. Cyclamen aldehyde, Lilial® and Bourgeonal® are the most fruity of the samples and are grouped together on the right-hand side of the map, while Lyral® and

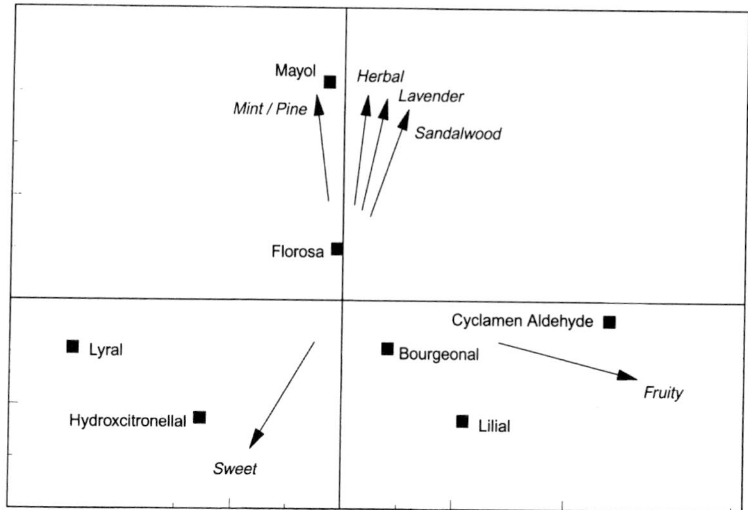

Figure 8.4 *Multidimensional scaling map of muguet perfumery materials*

hydroxycitronellal are the sweetest materials and Mayol® is the most herbal.

PSYCHOLOGY OF PERFUME

Olfaction is a powerfully emotive sense. An odour has the ability to remind us of the past; for instance, to bring back memories of situations from our youth, or to remind us of friends or family. It also has the ability to promote feelings of relaxation or comfort, a property which is made much of by aromatherapists, who also use it to aid the therapeutic properties of massage.

The process by which olfactory messages are interpreted is not yet fully understood, but it is known that olfactory messages are transmitted from the olfactory bulb along the olfactory nerve directly to the brain, where the path of the message divides into two. One route passes into the olfactory cortex at the front of the brain where identification and differentiation between odours occurs; the other passes into the limbic system at the centre of the brain. The limbic system is believed to be the emotional centre of the brain and it is here that many sensory messages are received and interpreted.

It is believed that this close link between the olfactive sense and the limbic region is the reason for such a close association between smell and emotion. To try to understand this link, researchers have studied the workings of the brain itself. Until recently, few data were available that characterized the nature of odour processing in the brain centres higher than the olfactory bulb. The problem lay in the complexity of the higher brain structures, the lack of understanding of brain mechanisms and the difficulty of detecting and locating brain activity. There is now a variety of imaging techniques that have been developed to investigate the structure and functionality of the brain.

Brain imaging methods were primarily developed for two purposes: firstly, to visualize structural information about the brain, and secondly to measure its functioning. The methods have developed to facilitate clinical knowledge, diagnosis and treatment and each has its limitations when applied to functional research studies such as olfaction. The chief methods used for visualizing the living brain include standard radiographic methods, contrast radiography, computerized axial tomography, magnetic resonance imaging, positron emission tomography and electroencephalography (EEG).

At Quest, we use spontaneous EEG to measure electrical activity of the brain from the surface of the scalp. A widely available technique, this is also extremely resource-intensive; a typical clinical session may take several hours to complete, including the preparation of the subject, the EEG experiment and collation of the resultant data. The vast amounts of data recorded take time to digitize, summarize and analyse.

In our work we have found significant differences in the quantitative and topographic changes in brain activity recorded from the scalp following presentation of a range of odour types, and relationships have been found between specific features of the recorded signals and measurable effects of the same stimuli on moods or feelings.

THE BUSINESS SCENTS BRIEF

The brief from Business Scents Ltd requests submissions for a feminine luxury line that includes a fine fragrance, antiperspirant, shampoo/shower gel and soap. The perfumer has asked for guidelines for the appropriate odour area as soon as possible so that the creative work can begin.

Definition of the Appropriate Odour Area

We already have information available from a project completed recently reviewing the perfumes currently used in toilet soap bars in Europe. A range of soaps was profiled by the sensory panel and the results were analysed to give a two-dimensional map showing the

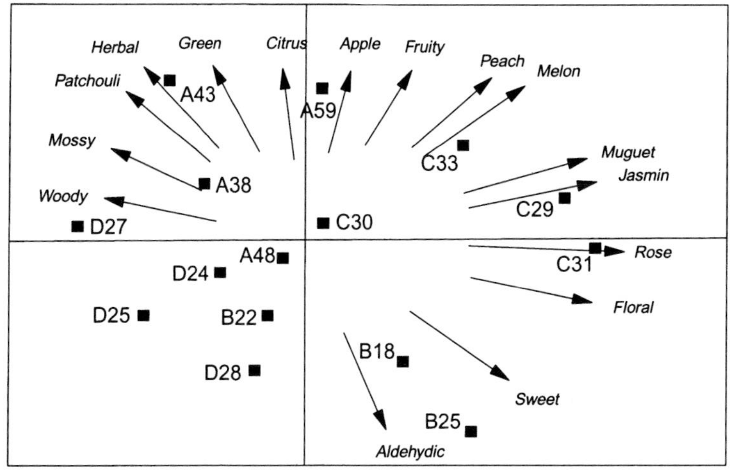

Figure 8.5 *Odour map of European toilet-soap bars*

relative similarities in odour character of the perfumes tested (Figure 8.5).

Those fragrances that are closest together on the map are most similar in odour character, whilst those that are furthest apart are most different. So, for instance, fragrances A59 and C33 are relatively similar in odour character, while A59 and D28 are quite different. To interpret the map and describe the nature of the differences between the odours, a correlation analysis is carried out. This analysis enables us to identify the characteristics that are most important in distinguishing between the fragrances, and the direction of increasing perception of each of these characteristics is indicated on the map with an arrow. So we can tell, for instance, that A59 is perceived to be far more fruity than D28.

We used a similar type of analysis to look for correlations between the odour character of the fragrances (defined by their position on the map) and results obtained from large-scale market research. The market research attributes found to be directly related to the odour perception of the samples are shown on the map in Figure 8.6, with arrows indicating the direction of increasing perception of each attribute.

The map indicates that the fragrances in the herbal, green and citrus areas are totally inappropriate for fragrance development for this brief: they are perceived by the consumer as functional, cleaning, invigorating and refreshing. Neither is the sweet floral, aldehydic area appropriate as, although it is seen as feminine, it is also perceived as cosmetic

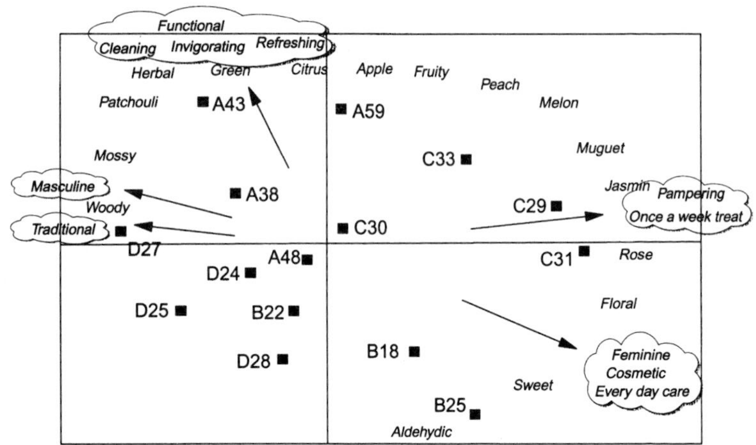

Figure 8.6 *Odour map of European toilet-soap bars with market research overlaid.*

or suitable for everyday care. The area that the perfumer needs to concentrate on is that described as white floral (muguet, jasmin) and fruity (melon, peach), which is perceived to convey the impression of a product that pampers you, and would be used as a once-a-week treat.

ANALYSIS OF INGREDIENTS TO CONVEY THE APPROPRIATE ODOUR CHARACTERISTICS

In the sections above, the sensory profiling and multidimensional scaling techniques are illustrated using data obtained from the analysis of muguet perfumery materials. From a purely olfactory point of view, the type of odour that best fits the odour described in this brief is cyclamen aldehyde, but with the range of products that need to be considered in this brief the choice of material is not quite so simple.

The chemistry of the stability of fragrance materials in different bases is discussed in Chapter 9. The olfactory implications of these changes can also be characterized using a sensory test similar to that carried out with the muguet materials, but placing each at appropriate concentrations (or range of concentrations) in the target-product bases.

CREATING FRAGRANCES FOR THE FUTURE

As we approach the twenty-first century, theories are being built and expanded upon that are almost self-fulfilling in creating the trends of the future in fashion, art, family values, *etc*. As fashion designers, social philosophers and economists design our clothes and way of life for the future, so perfume designers seek to create the smells that complement and reinforce this environment. At the same time, a new and exciting smell within the constraints of acceptability gives the designer the advantage of novelty so sought after in the industry.

With the powerful software currently available, it is possible to pull together information from a wide variety of sources to give the perfumer guidelines to help channel or expand creative ideas in a way that is calculated to produce a successful odour type. Traditionally, market research gathers information from consumers in retrospect, using perfumes that have been selected as appropriate for a product and measuring a response to those perfumes, which is inevitably based on experience. For a new trend-setting perfume style, we need to predict or create a popular perfume trend for the future.

A perfume company such as Quest International establishes a perfume and product profile based on large studies of consumer responses (acquired from consumers of different age groups and social profiles) to different perfumes in different products in different countries over a number of years. At the same time, a profile is built of changing social patterns and how this affects acceptance of different perfume styles. As part of our on-going studies of evaluative responses to odour types a profile is also built of the feelings or images that are conveyed by different odours.

The availability of evaluative and market research data, social trends analysis and fashion information, and perfume stability and olfactory performance measures, all of which are structured within the odour analysis provided by the sensory analysis group, provides the perfumer with a basis from which to work and add creative skill to develop the new fashion fragrance for the year 2000.

Chapter 9

The Application of Fragrance

JUDI BEERLING

THE ROLE OF AN APPLICATIONS DEPARTMENT

An Applications Department within a fragrance house is responsible for 'applying' a fragrance oil or essence to a consumer product. This process in itself can be problematic, since the fragrance may not always be easily incorporated into the product base. Solubility of the fragrance could be a problem, for example, in an aqueous fabric conditioner system, a volatile, silicone-fluid-based antiperspirant or a hair dressing based on mineral oil. This might manifest itself as a haziness, as floating droplets of fragrance or as complete separation of a clear or milky layer. Dosing of liquid fragrance into talc or powder detergent can prove difficult if perfume dosage levels are above the norm, affecting the powder's flow characteristics or causing lumping.

However, the job does not stop there, even once the perfume has been successfully dosed into the product. A number of other factors have to be studied to ensure that the perfume remains evenly dispersed within the product and does not cause its physical and chemical characteristics to change significantly over time. The integrity of the product needs to be checked over a period of accelerated storage at different conditions. The latter are often stipulated by the product manufacturer, who wants to ensure that the product's characteristics or activity are unaltered by the fragrance, even if exported to tropical climates or sub-Arctic conditions. However, equally important to the perfumer who is designing a new fragrance, or an evaluator who is selecting a perfume from a repertoire or 'shelf', is to monitor any changes in the fragrance odour once in the product base, for example:

—Fragrance intensity, *i.e.* does it still have good impact?
—Perfume character, *i.e.* does the perfume still smell the same or very close to that of a newly made product (allowing for some desirable maturation or ageing of the perfume in the product over a 1 or 2 week period)?

The whole subject of stability testing is discussed in more detail later in this chapter.

Thus, it is the job of the applications chemist to be the 'product expert', who should be able to advise and carry out testing on any potential problems or issues related to the perfume in combination with the product base. This advice may be sought by:

—Perfumers and evaluators internally, to guide their choice of perfume types and individual raw materials suitable for a particular product formulation.
—Sales personnel, to enable them to ask the right questions of their clients, *e.g.* what active ingredients might the product contain or is the type of packaging the client will use likely to cause a problem?
—Marketing personnel, of both the supplier and client, to help them decide on technical product-concept feasibility and the likely constraints on fragrance creativity.
—The client's technical or R&D department, to discuss likely test protocols, advise on starting-point product formulations or possible product and perfume interactions to be taken into account during their development work or factory manufacture.

It is also true to say that part of the Applications Department's role is to service any kind of client request for technical information, ranging from where a particular product raw material might be purchased to what sort of natural or functional additives could be added to the product to convey a particular consumer benefit.

To illustrate how an Applications Department works, the approach to the Business Scents Ltd brief given in Chapter 6 is discussed.

PRODUCT FORMULATIONS

Before work can begin on sampling or testing any of the perfumer's new creations, the client has asked for suggested starting-point product formulations. Business Scents Ltd does not have the necessary development resource or manufacturing equipment in-house to handle the production of any of the products other than the alcoholic fine fragrance, and will appoint contract manufacturers to do this at a later date. Prototype product samples that contain experimental fragrance creations are thus required for the client to hold initial focus group discussions with target consumers. The applications chemist assigned to the team handling this project therefore needs to be able to produce base formulations within the first few days or weeks, in order for the perfumer's initial creations to be incorporated and the evaluation process to begin. In this case, the client has requested quite conventional, well known products and suitable examples are available from the bank or repertoire of existing tested and approved (in terms of odour, colour and physical characteristics) formulations. These formulations are sent to Business Scents Ltd and, if they meet their requirements, initial batches of the unfragranced product bases can be prepared in the applications laboratory.

Fine Fragrance

The Business Scents *Eau de Parfum* formulation (Figure 9.1) contains 78% denatured ethanol. The source of this ethanol (*e.g.* synthetic; or natural grain starch, sugar beet, or molasses alcohol) can give rise to a different odour in the end product, and thus it is important to know what type of alcohol is likely to be used. Denaturants (which deter

Formulation

Ingredient	*% w/w*
Ethanol (DEB 100)	78.00
Fragrance	12.00
Purified water	8.50
PPG-20 methyl glucose ether[a]	1.00
Benzophenone-2[b]	0.50

Preparation
Blend the fragrance and ethanol, then add the water and remaining ingredients slowly with mixing. Allow to mature at room temperature for up to 10 days and then cool to +1 °C, followed by filtration. A filter aid, such as magnesium carbonate at 0.2% can be used to remove difficult precipitates. Fill into clean, glass bottles.

Figure 9.1 Eau de Parfum *formulation ([a] Glucam P20® ex Amerchol Corp, Edison, NJ, USA; [b] Uvinul D50® ex BASF AG, Ludwigshafen, Germany)*

people from ingesting the alcohol) are required by legislation in many countries and are also useful to know, although less likely to have a significant effect on the finished product. In this case, a standard 99.7% v/v synthetic grade, containing 0.1% v/v *t*-butanol as a marker and denatured with 10 p.p.m. of Bitrex [INCI[a] name is denatonium benzoate, an extremely bitter substance], is likely to be used by the client. Given the high level of alcohol (and consequent low level of water), there is unlikely to be any need for addition of solubilizers.

The formulation also contains an ultraviolet (UV) radiation absorber, benzophenone-2, to prevent degradation of the fragrance and any dyes by light. Although consumers are encouraged to keep fine fragrances in the dark, the manufacturer needs to protect the product from those of its customers who insist on storing it on a sunny windowsill! A moisturizing ingredient which has additional fragrance fixative properties (PPG-20 methyl glucose ether) is also incorporated in this instance.

As stated in Chapter 7, the perfumer has almost total freedom in creating perfumes for such products and the Applications Department simply needs to check the solubility of the fragrance at a range of temperatures, following the maturation, chilling and filtering process. If the maturation period is sufficient, there should be very little chance of any further solid materials in the fragrance precipitating out over time, although this must be checked. Light stability, even though a UV absorber is being used, must also be checked to ensure that the fragrance does not darken unacceptably or that any dyes added do not fade. It may be that, in this case, a different UV absorber works better or that the perfumer needs to change one or two of the ingredients in the fragrance.

Vegetable Soap

The client has asked for a recommended formulation for a luxurious soap bar, based solely on vegetable fats (Figure 9.2). A high-quality palm–coconut base that is widely available in Europe should be suitable, but it was felt that such a premium product should contain some additional materials to convey 'added value'. Thus, a powdered cationic polymer is incorporated, which deposits a moisturizing film on the skin and helps to promote a rich, creamy lather. Titanium dioxide, a standard additive to any soap, provides greater opacity (and, to some extent, whiteness) to the bar and a variety of pigment pastes, which are

[a] International Nomenclature for Cosmetic Ingredients (INCI) names are used throughout, since this is the standard recognized in the industry.

Formulation

Ingredient	*% w/w*
Prisavon 9259® soap base[a]	To 100.0
Merquat 2200®[b]	1.00
Fragrance	1.70
Titanium dioxide	0.20
Colourant or white slurry	As required

Preparation
The Merquat®, titanium dioxide, colorant and fragrance are mixed with the soap base, which has been milled once previously. This mix is then milled three further times, followed by plodding, extruding and stamping into bars.

Figure 9.2 *Conditioning vegetabe soap formulation ([a] 80:20 palm:coconut soap base, ex Unichema Int., Gouda, Netherlands; [b] polyquaternium-7, ex Chemviron Speciality Chems., Overijse, Belgium)*

stable in the alkaline environment (around pH 9–10) and to light, are available to colour the soap any desired shade. A white soap also contains colourant (called a white slurry, consisting of a small amount of a blue pigment and a fluorescent whitening agent) because raw soap base is, in fact, a dull, yellowish-cream colour.

What sort of stability issues are there likely to be in such a soap formulation? Firstly, the colour of the fragranced soap is as yet undecided; if it is to be white, then the degree of discoloration which can be tolerated is likely to be a lot less than if the bar is coloured to 'match' the fragrance type. The disastrous consequences of incorporating too much vanillin into a fragrance for white soap are shown in Figure 9.3. This severe browning reaction begins to occur after a matter of hours and cannot be prevented, since it results from a chemical reaction caused by the high pH of the soap, which is accelerated by exposure to light. For this reason, vanillin is rarely a major ingredient in soap perfumes. It may be used in small amounts in a soap if it is known that it is to be dyed a strong colour. There are alternative materials with vanilla-type odours that can be used instead, such as ethyl vanillin (which, although still discoloring, can be used at a lower level), Ultravanil® and Benzoin Hypersoluble P85®. Other perfumery materials that can cause a less radical, but still potentially significant, change in the colour of a white bar include eugenol, isoeugenol, heliotropin, certain mosses, Schiff's bases, citral, indole, *etc.* These

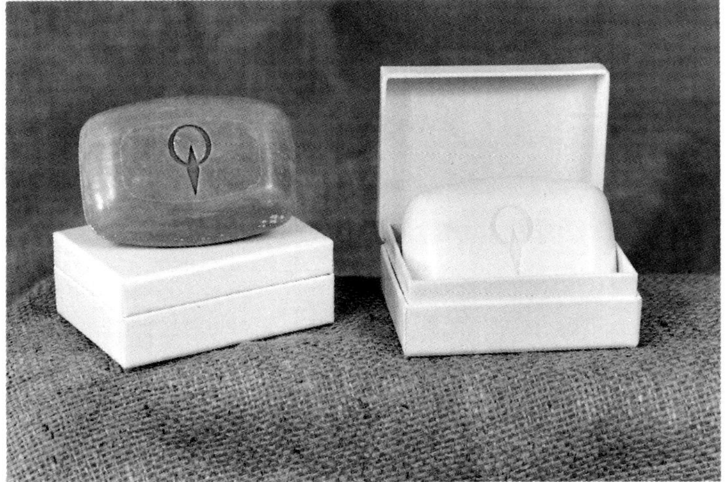

Figure 9.3 *The effect of vanillin on white soap*

also need to be avoided, substituted or used only very sparingly and colour changes measured to ensure they stay within specification, even after storage at high temperature (see later).

Aerosol Antiperspirant

The client requires an effective antiperspirant product which is to be dispensed as an aerosol spray using a propane–butane blend as the propellent gas. Figure 9.4 illustrates the basic components of an aerosol.

The formulation that the applications chemist uses (Figure 9.5) is a powder in volatile silicone-fluid (cyclomethicone) suspension. This type of formulation requires the use of a special valve and actuator system, which allows the powder active (activated aluminium chlorhydrate) to be dispensed without clogging. The antiperspirant active chosen gives good sweat reduction, and has a suitable particle-size distribution which can be effectively dispensed through the valve and actuator. The cyclomethicone carrier fluid is sufficiently volatile to evaporate from the skin surface and is soluble in the propellant blend, giving a suspension system that is easily dispensed from the can. A high molecular weight silicone gum (diluted in cyclomethicone) is also included to help prevent the formation of an aerosol 'cloud', which can cause choking or sneezing if inhaled. It also provides a soft,

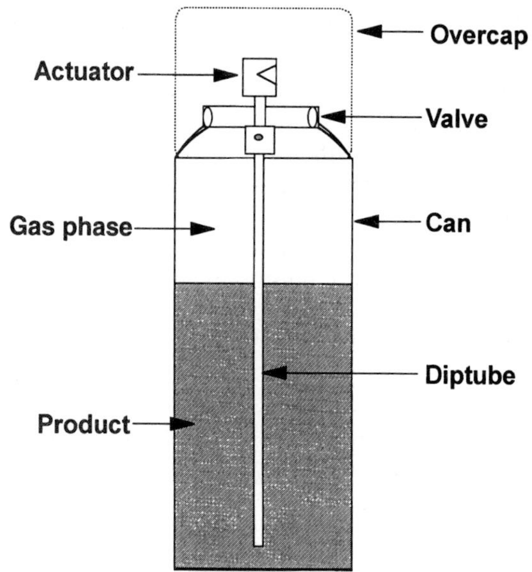

Figure 9.4 *The basic components of an aerosol*

Formulation

Ingredient	*% w/w*
Concentrate	
Isopropyl myristate	10.00
Bentonite	3.65
Volatile silicone-fluid DC 344[a]	48.65
Activated aluminium chlorhydrate powder	36.00
Dow Corning® 1401[b]	0.50
Fragrance	1.20
Aerosol fill	
Concentrate	25.00
Propane–butane propellant, 40 psig	75.00

Preparation

Thoroughly disperse the bentonite and activated aluminium chlorhydrate in the isopropyl myristate and DC 344 fluid, using a high-shear mixer. Add the DC 1401 and the fragrance, mixing well. Fill into internally lacquered aluminium aerosols, as shown above.

Figure 9.5 *Dry aerosol antiperspirant formulation ([a] cyclomethicone, ex Dow Corning Int., Brussels, Belgium; [b] cyclomethicone and dimethiconol, ex Dow Corning)*

moisturized skin feel, which is important for the image of such a prestige product.

What sort of problems can an aerosol antiperspirant cause the perfumer and the applications chemist? Empirical testing in a range of antiperspirant bases has identified a number of unstable perfumery ingredients, such as phenolic materials, unsaturated terpene alcohols and their esters (such as linalool, linalyl acetate), unsaturated or reactive aldehydes (such as Lilial® and Ligustral®) some essential oils (such as bergamot, lavender and rosemary), citrus oils, and spices, (such as nutmeg and black pepper), that can undergo chemical reactions in the presence of the acidic antiperspirant active. Such materials should be avoided. Being an aerosol product does in itself create a challenge; traces of metal ions can cause the active or perfume to discolour to pink or brown, which would not be very attractive under the arm or on clothes!

Solubility of the fragrance in the volatile silicone and the propane–butane, although not visible to the consumer, might cause a change in note of the perfume as it is dispensed. The potential for the product to cause can corrosion must also be evaluated. This can be tested by storage testing of the finished product in production specification cans. It is rare for modern anhydrous systems to present major corrosion problems, although all formulations should be checked, especially if they contain significant levels of water.

In any type of product, and especially aerosols, the balance between how the different components of the fragrance are perceived, the so-called odour balance, is altered compared with the neat fragrance oil. Very generally, in an aerosol system, the increased surface area afforded by the droplets of the aerosol spray, combined with the highly volatile propellant, tend to give a greater bias towards the more volatile components of the fragrance compound. This causes the initial fragrance perception to be biased towards the top-notes. Such a change in odour balance is especially marked in a product such as the anhydrous antiperspirant, with its high level of propellent. This is why all aerosol fragrances should be evaluated with a formulation and valve system as close as possible to the one used in the finished product. It is also a factor that the perfumer needs to consider when creating the 'Eve' fragrance modifications for this product.

In conclusion, this product is probably the most aggressive and potentially complex product of the five to perfume. It is highly likely that the fine fragrance formulation will need to be significantly modified to be stable in this medium, whilst still giving the overall impression (in use) that it has essentially the same smell.

Shampoo

Figure 9.6 shows the conditioning shampoo formulation selected. It is based on a blend of a high foaming anionic detergent (sodium lauryl ether sulfate, commonly abbreviated to SLES) and a mild, amphoteric betaine detergent which lowers the irritation potential of the SLES. A cold-mix pearl concentrate is added to give an attractive, white, pearly appearance to the product. Conditioning is provided by several keratin substantive materials, which deposit a thin protective film on the hair, smoothing down the cuticle and preventing static build-up. A copolymer of wheat protein/silicone glycol is used, since evidence from the manufacturer shows that this helps to shield the hair from environmental damage. Panthenol (commonly called Pro-Vitamin B5 on many leading hair care products) is added to thicken and help repair damaged hair. The final ingredients are preservative (the final choice of which is made by the manufacturer) to prevent microbial spoilage of the product, sodium chloride to thicken the detergent system and, of course, water (which must be deionized and substantially free from bacteria).

The last ingredient, although seemingly innocuous, can cause major problems for the perfumer. In Chapter 7, the perfumer discussed the

Formulation

Ingredient	*% w/w*
Purified water	to 100.00
Sodium lauryl ether sulfate, 28%	41.50
Dehyton K®[a]	8.00
Panthenol	1.00
Crodasone W®[b]	3.00
Preservative	As required
Euperlan PK-3000®[c]	3.00
Fragrance	0.60
Sodium chloride	2.00

Preparation
Add the ingredients to the water in the order given, with efficient stirring. (The level of sodium chloride may require adjustment.)

Figure 9.6 *Frequent-use conditioning shampoo formulation ([a] cocamidopropyl betaine, ex Henkel Chemicals, Dusseldorf, Germany; [b] hydrolysed wheat protein polysiloxane copolymer, ex Croda Chems, Goole, UK; [c] glycol distearate and laureth-4 and cocamidopropyl betaine, ex Henkel)*

effect that the polarity, or relative water solubility of the perfumery ingredient versus its molecular mass, has on its odour perceptibility from an aqueous detergent system. The nature, quality and level of the detergent(s) and other materials, such as protein conditioners, can sometimes dramatically affect the perfume headspace over the surface of the product and its in-use odour characteristics. The perfumer thus asks the Applications Department to try various fragrance modifications in the detergent base, to assess on smelling the product straight from the bottle and possibly also during a salon evaluation in which a panel of volunteers has the odour of their hair assessed during washing and immediately after rinsing. Equally important is to monitor the smell on washed and dried hair switches (small samples of untreated, virgin hair) for up to 18 hours afterwards, if substantivity or malodour counteractancy is important.

Another problem that the applications chemist may have to contend with is that the perfume may cause shampoo viscosity to decrease or, more often, increase when added to the unfragranced base. If this is a small change, it can be accommodated by varying the level of thickener. However, if dealing with a client's fully formulated base, this may not be an easy option. In such cases, the perfume formulation needs to be screened for ingredients that are known to affect the viscosity of surfactant systems, such as dipropylene glycol or alcohols, *e.g.* citronellol, which often causes a decrease, and diethyl phthalate, isopropyl myristate or terpenes, which can thicken such solutions quite dramatically.

Obviously, the exact formulation of the client's shampoo under consideration may be unknown to the perfumery house and not follow any previously experienced pattern. If so, it may be necessary to resort to single-ingredient testing of perfumery materials to discover which ones can be safely used without problems. Similarly, chemical reactions could take place between certain perfumery raw materials and active ingredients in a shampoo, such as an anti-dandruff agent like piroctone olamine. Such reactions could lead to discoloration or off-odours in the product, which may only be noticeable after a period of high-temperature storage. As it is a previously tested shampoo base with no problem actives, the formulation given in Figure 9.6 is not expected to give rise to major difficulties but, nevertheless, standard stability testing is initiated when the final fragrances have been selected for submission, to avoid any unpleasant surprises. This is carried out in glass jars, as well as Business Scents Ltd bottles, if available at this stage. Compatibility problems are unlikely to be experienced with polyethylene terephthalate (PET), although it could be affected by citrus and terpene materials. Since the client has specified clear packaging, light testing is also necessary (see later).

Shower and Bath Gel

The basic formulation used for this product is given in Figure 9.7. It is based on a mild blend of a high-foaming SLES, as for the shampoo, plus a high-purity betaine surfactant, which also contains glyceryl laurate to help build viscosity. The active detergent level (18%) is higher than that of the shampoo (14%), because of the greater dilution factor that the shower and bath gel will experience in use. A mild, vegetable-derived humectant, which moisturizes and softens the skin, is included, along with a cationic skin conditioner (Necon CPS-10O®), which can be perceived on the skin even after dilution in the large volume of water used in a bath. Additional thickening to the desired

Formulation

Ingredient	*% w/w*
Phase A	
Purified water	to 100.00
Sodium lauryl ether sulfate, 28%	54.00
Tegobetain HS®[a]	8.00
Phase B	
Purified water	20.00
Antil 171®[b]	3.00
Preservative	As required
Gluquat 125®[c]	6.00
Necon CPS 100®[d]	2.00
Dow Corning® 193 surfactant[e]	0.70
Phase C	
Fragrance	1.00

Preparation

Blend together the ingredients of phases A and B separately. Slowly add phase B to A, with moderate agitation, until homogeneous. Finally, add the fragrance and mix well.

Figure 9.7 *Shower and bath gel formulation ([a] cocamidopropyl betaine and laureth-4, ex Th Goldschmidt AG, Essen, Germany; [b] propylene glycol and PEG-55 propylene glycol oleate, ex Th Goldschmidt; [c] lauryl methyl gluceth-10 hydroxypropyl dimonium chloride, ex Amerchol Corp., NJ, USA; [d] PEG-15 cocamine phosphate/oleate complex; [e] dimethicone copolyol, ex Dow Corning Int., Brussels, Belgium)*

Table 9.1 *Botanical extracts with 'relaxing' connotations*

Botanical extract	*Claimed benefit*
Passion flower (*Passiflora incarnata*)	Reported to have analgesic and sedative properties; used in sedative preparations for nervous anxiety
Hawthorn (*Crataegus laevigata*)	Flowering tops are used in sleep-inducing preparations; reduces blood pressure
Valerian root (*Valeriana officinalis*)	Hypotensive; widely used in sleep aid and sedative health foods, herbal teas and medicinal products
Linden (lime tree; *Tilea cordata*)	Sedative, emollient, anti-inflammatory, healing and refreshing; linden tea or 'tilleul' is drunk as a general relaxant

gel-like consistency is achieved with a proprietary liquid thickening agent, Antil 171®.

Being an aqueous surfactant system, similar issues to those for shampoo need to be considered. The major difference is that the shower and bath gel is designed to be crystal clear and a thick, but still mobile, easily dispensed gel. Thus, the solubility characteristics of the fragrance are critical, as are any significant viscosity variations. The client has asked for recommendations for suitable botanical extracts which could support a relaxing claim for the product. Such extracts often contain low levels of essential oils, which could add an odour of their own to the base and may reduce the viscosity by virtue of containing propylene glycol as a carrier or solvent. Thus, it is useful to add these ingredients to the finished product base to ensure that the fragranced product still performs well. Table 9.1 gives examples of possible additives that might be recommended.

Both of the detergent products will be coloured with water-soluble dyestuffs to enhance the fragrance concept. Blue, green or peachy–orange are all possible choices for the colour, but unfortunately both blue and red dyes, in particular, are notoriously unstable to light. Thus, as with the shampoo, light-stability testing is needed to ensure that ingredients in the perfume do not destabilize the colour of the product, since it is likely to be packaged in clear or translucent containers.

STABILITY TESTING

As touched on previously, it is seemingly impossible to predict accurately, in the few short weeks or months of the average product-development process, the chemical and physical changes that might occur after anything up to several years of storage on a consumer's bathroom shelf, or in the dusty corner of a drug store in Timbuktoo. Similarly, the extremes of temperature and/or humidity that a product can be exposed to during transportation or sitting on a quayside in certain parts of the world can be severe, even if only for relatively short periods. Thus, over the years, accelerated test protocols have been developed to provide the manufacturer with a way of predicting likely shelf-life (which is normally a minimum of 2 years) and identifying potential problems before the product reaches the marketplace. The applications chemist has adapted these protocols to highlight any undesirable chemical or physical changes related to the fragrance. Thus, before submission to Business Scents Ltd, the leading contenders for the 'Eve' fragrance and its line extension modifications are submitted to a battery of stability tests, as alluded to previously.

The Arrhenius rate equation forms the basis for the theory behind this accelerated testing. This states that for every 10 °C increase in temperature, the rate of reaction doubles. Thus, in theory, the following applies:

$$
\begin{aligned}
12 \text{ months at } 20\,^\circ\text{C} &= 12 \text{ weeks at } 40\,^\circ\text{C} \\
&= 6 \text{ weeks at } 50\,^\circ\text{C} \\
&= 3 \text{ weeks at } 60\,^\circ\text{C}
\end{aligned}
$$

In practice, perfumery companies test all products at 0–4 °C, 20 or 25 °C and 37 °C (or 40 °C) for 12 weeks as a minimum standard. In addition, tests at 45 or 50 °C may be used (except for aerosols, which could explode) if quick results are required, or if it is likely that the product will experience these sort of conditions during its life-cycle. Stability for 4 weeks at 50 °C is considered acceptable in many instances, but signs of instability should be taken as precautionary only, particularly if the samples still appear satisfactory at 37 or 40 °C. This is because certain chemical reactions could occur at these high temperatures that could not happen at ambient temperatures or even at 37 °C. Note that 37 °C has become an accepted standard because many cosmetic chemists believe there is a temperature 'barrier', corresponding to the heat of the human body, above which chemical reactions are accelerated beyond that predicted by the Arrhenius equation.

Humidity Testing

Products that are likely to be packaged in permeable materials, such as paper or cardboard, need to be tested in high-humidity conditions.

Thus, a common combination is 37 °C/70% relative humidity (RH) or 40 °C/80% RH. This type of testing is most useful for boxed laundry powders (particularly those destined to be sold in tropical climates), but may be important for the 'Eve' soap if it is to be sold in a wrapper or carton rather than a tightly sealed pack.

Light Testing

A great deal of contention surrounds how best to accelerate the testing of products that will be exposed to daylight, and particularly to strong sunlight. Prior to the development of suitable equipment, samples were placed on a North-facing (so that they were not in direct sunlight) windowsill for however long was available for testing. However, quite variable results are obtained, depending on geographical location and season. Although such tests are often still conducted as a back-up, the availability of UV cabinets, such as the Hereaus Novasol Test®, has helped to standardize testing. The Xenon arc lamp subjects the samples to UV light in the 300–800 nm wavelength range and can run with a 400 or 1000 W burner fitted. Exposure of 6 h to the 1000 W lamp is usually sufficient to see any changes that are likely to occur in about 3 months of daylight testing. There are, however, considerable disadvantages to this method, including the fact that, despite the presence of a cooling fan, the temperature inside the cabinet can become quite hot, and thus two variables, rather than one, are being tested at once. Also, UV light is not a true reflection of normal daylight and it is possible for some discoloration reactions to occur in the first few hours and then bleach out again before the 6 h is complete. Thus, some companies now invest in new light stress chambers, which can be fully temperature controlled (between +10 and +50 °C) and can be run with simulated daylight bulbs (for real-time testing), UVA or UVB tubes or fluorescent light (to simulate in-store conditions).

Assessment and Reporting of Results

After 12 weeks of stability testing has been completed (and usually after an interim 4- or 6-week period), all the perfumed samples are assessed on odour and physical appearance. At Quest, scales of A–E for perfume character and 1–5 for odour intensity are used. The refrigerated sample (nominally 0 °C) is treated as a standard, as perfume degradation is considered to be negligible at this temperature and is therefore rated as Al. The samples from the other conditions are assessed and rated against this standard. Any perfumed sample rated below C3 is regarded as unsatisfactory.

On physical appearance, the samples are checked for discoloration, separation, precipitation, changes in viscosity, *etc.* as appropriate to the product. A viewing cabinet can be useful for looking at the samples under consistent lighting conditions. An unperfumed sample or 'control' is always tested alongside to compare with the perfumed products. An A–D scale is used, in which C or D are considered a failure.

Most products are assessed directly from the bottle or jar, soap wrapper or box. However, some products, the true odour characteristics of which show up better in solution, are dissolved in warm water in an assessment cup. Aerosols are assessed on test pads, with time

Table 9.2 *Stability test results for the Eve Project (comments would be given on any significant changes in the odour or physical appearance, and results for the unperfumed controls have been omitted for simplification)*

	Odour				*Physical appearance*				
	0 °C	20 °C	37 °C	50 °C (4 weeks)	0 °C	20 °C	37 °C	50 °C (4 weeks)	UV (6 h)
'Eve' *Eau de Parfum*	A1	A1	A/B2	A/B2	A	A	A/B	B	B
Soap Version 1	A1	B2	C3	C3/4	A	B	C	D	D
Soap Version 2	A1	A2	B2	B2	A	A	A/B	B	A/B
Aerosol A/P Version 1	A1	C2	C3	–	A	B/C	D	–	–
Aerosol A/P Version 2	A1	A1	A2	–	A	A	A	–	–
Aerosol A/P Version 3	A1	A2	B2	–	A	A	A/B	–	–
Shampoo/ Shower gel Version 1	A1	B1	B2	B/C2	A	A	A	B	C
Shampoo/ Shower gel Version 2	A1	A1	B1	B1	A	A	B	B/C	A

Odour character: A = no change, excellent, B = very slight modification, C = some modification but acceptable, D = modified, unsatisfactory, E = unrecognizable; Odour intensity: 1 = no change, 2 = slight weakening, 3 = some loss of impact but acceptable, 4 = weak, unsatisfactory, 5 = odourless or extremely weak; Physical appearance: A = no change, B = slight change, C = changed but acceptable, D = unacceptable change, *e.g.* discoloration, separation, viscosity change, *etc.*

allowed for the propellant to evaporate. Testing in special, plastic-coated, clear, glass aerosols may be required in cases where solubility or discoloration reactions need to be observed inside the product.

The results for the project 'Eve' stability tests are shown in Table 9.2. These are reported to the perfumer and evaluator assigned to the project, and then are fed into the stability database on the computer for future reference.

Thus, it can be seen that several of the fragrance modifications created for the line extensions have successfully completed the standard stability tests and the final selection can be made based on additional criteria, such as cost and hedonics. These fragrances can then be submitted to Business Scents Ltd with confidence that they are likely to be stable for at least a year, as long as the bases used are not radically altered. However, it remains the client's and their contract manufacturer's responsibility to carry out their own stability testing in the final product formulations to ensure their consumer safety and acceptability.

BIBLIOGRAPHY

H. Butler, *Poucher's Perfumes, Cosmetics and Soaps*, Vol. 3 Cosmetics, 9th edn, Chapman & Hall, London, 1993.

G. Holzner, *Cosmet. & Perfum.*, 1974, **89**, 37.

IFSCC Monograph No. 2, *The Fundamentals of Stability Testing*, Micelle Press, Weymouth, 1992.

J. Knowlton and S. Pearce, *Handbook of Cosmetic Science and Technology*, Elsevier Advanced Technology, Oxford, 1993.

A. Leung and S. Foster, *Encyclopedia of Common Natural Ingredients used in Food, Drugs and Cosmetics*, 2nd edn, John Wiley & Sons, New York, 1996.

Chapter 10

The Safety and Toxicology of Fragrances

STEVE MEAKINS

INTRODUCTION

How many consumers, even the more discerning ones, ever think about the safety of a fine fragrance or aftershave before they buy it? It is likely that they are far more interested in ensuring that the product 'smells right' on their skin, and that it blends with and enhances the image they are trying to establish. When 'Eve' is launched, its image will be the major selling point, not the fact that it is safe for the consumer to use. However, it is certainly not by chance that the consumer does not consciously think about fragrance safety when selecting a product. Their confidence comes from the long history of safe use associated with fragrances, which in turn is the result of the considerable effort that the fragrance industry applies to product safety. The fragrance industry is actually 'self-regulating' as there is no legislation, apart from normal consumer laws, that govern the composition or use of fragrances.

SELF-REGULATION

It was realized by the fragrance industry some 30 years ago that this absence of regulations concerning the ingredients that could be used in fragrances, or on the safety of fragrances in consumer products, could expose the consumer to unacceptable risks which would lead to governmental intervention in the industry. To avoid such problems, the industry decided to establish a self-regulatory system involving the two major international fragrance organizations: the Research Institute for Fragrance Materials (RIFM) and the International Fragrance Association (IFRA).

RIFM was established in 1966 by the American Fragrance Manufacturing Association as a non-profit-making, independent body, whose task was to evaluate the safety of fragrance ingredients. To date, RIFM has tested over 1300 fragrance materials, including all of the commonly used ingredients. Once the test results for each material examined have been reviewed and discussed by an independent international panel from academia, which comprises toxicologists, pharmacologists and dermatologists, the results are published as monographs in the journal *Food and Chemical Toxicology*. RIFM also collates all the information available for an ingredient from the scientific literature and from the aroma chemical manufacturers for inclusion in these monographs. Should there be any cause for concern about the use of an ingredient, this is immediately signalled to the industry by publication by RIFM of an advisory letter, which is then acted upon by IFRA.

The types of basic test carried out by RIFM include acute oral toxicity, acute dermal toxicity if the oral toxicity is significant, skin irritation and sensitization, and phototoxicity if the material adsorbs in the UV range. Where there is a need, much more detailed studies are undertaken which involve subchronic feeding studies, dermal absorption and metabolic fate. Through IFRA, RIFM also collects from the industry consumer exposure data on fragrance ingredients. This ensures that the test data it uses are relevant to the market situation and also provides guidance on the nature of future research. Thus, RIFM undertakes a review of its safety data, or instigates further research if the results of these surveys indicate that a particular ingredient is occurring in a wider range of products and/or at higher concentration than when it was first examined.

IFRA was established in 1973 by a number of fragrance trade associations and represents over 100 fragrance manufacturers in 15 countries. With its headquarters in Geneva, IFRA represents the scientific and technical expertise of the industry and is responsible for issuing and up-dating the *Code of Practice* (IFRA 1973) upon which the whole self-regulation policy is based. IFRA is funded by these fragrance manufacturing companies, who all agree to abide by the code of practice whilst they remain members of the association. This code of practice has many functions, including setting standards for good manufacturing practice within the industry, for quality control, for labelling and advertising, as well as setting limits on, or prohibiting the use of, certain ingredients.

Although IFRA and RIFM are independent of each other they obviously work closely together and it is only after considerable discussion between RIFM and IFRA that restrictions or prohibitions are imposed. It is always the IFRA board, in conjunction with the technical advisory committee, that makes the final decision in such matters, as the IFRA board is ultimately responsible for the implementation of any restrictions, by way of the code of practice.

The strict code of practice applied by IFRA not only protects the consumer, but also protects the health and well-being of those employed within the industry. This is highlighted by a study carried out in 1985 in which it was found that there was no increase in mortality from any type of cancer in a group of workers employed in the flavour and fragrance industry, where exposure to a wide range of aroma chemicals is far higher than the consumer would ever encounter (Guberan and Raymond, 1985).

As the fragrance industry does not use toxic, carcinogenic or corrosive substances, why are there use restrictions on some ingredients? The most common cause for such a restriction is the ability of some materials to be skin sensitizers. Unlike skin irritation, which usually disappears soon after the irritant has been removed, skin sensitization involves the activation of the immune system and reactions can persist for much longer after the initial exposure and can become more severe on subsequent contact.

Skin sensitization was recognized as a major problem almost at the very outset of ingredient testing and a simple strategy was devised to deal with it. Ingredients were tested in a human, predictive sensitization patch test at a concentration ten times greater than the consumer was likely to be exposed to. If this test (which used exaggerated levels under occluded patches as a 'worse case' scenario) proved to be negative, then no further action was taken. However, in the early days, if there were signs of sensitization, RIFM would issue an advisory letter and the material would no longer be used in fragrances.

As this approach did not differentiate between weak and strong sensitizers, IFRA subsequently adopted a modified approach. The IFRA technical advisory committee studied the results of the human patch tests, and if necessary asked for further work to be undertaken, to see if a 'no effect' level could be determined for each material. If such a level could be determined, the committee set a guideline value that allowed only one-tenth of the no effect level to be used in a consumer product. Obviously, in some cases this level was below that at which the ingredient made any useful contribution to the fragrance and IFRA recommended a complete ban. However, it did allow many useful ingredients, such as hydroxycitronellal (1), cinnamic alcohol (2) and isoeugenol (3) to be used without exposing the consumer to unnecessary risk. An interesting example of this approach is *trans*-hex-2-enal (4), which has an intense green, fruity, vegetable-like odour. This material was found to be a sensitizer at a level of 0.2%, but not at 0.02%. IFRA thus set a use level of 0.002% (20 p.p.m.) for a consumer product, at which level it can still have a strong influence on a fragrance, especially as its odour threshold has been measured at less than 0.1 p.p.m.

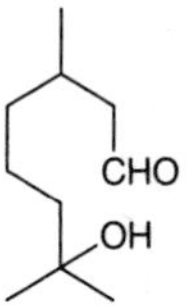

CHO
OH
Hydroxycitronellal
(1)

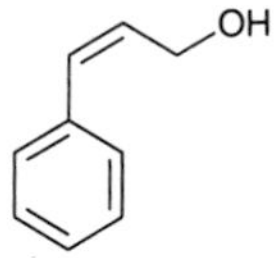

OH
Cinnamic alcohol
(2)

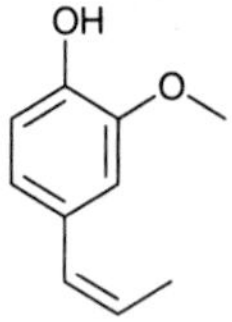

OH
O
Isoeugenol
(3)

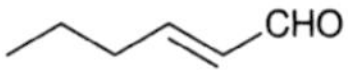

CHO
trans-Hex-2-enal
(4)

SAFETY ASSESSMENT

It was realized many years ago that the only difference between a medicine and a poison was the dose administered, as is demonstrated by the tragic consequences of overdosing on paracetamol. It is thus clear that it is not accurate to say a material is safe, as safety cannot be measured in absolute terms. The safety of any material cannot be measured directly, but can only be estimated as part of a risk assessment. This type of assessment examines the potential of any material to cause harm and the likelihood of that potential being reached during normal-use conditions. If the probability of causing harm under normal-use conditions is high, then the material is not used, or a different use-regime is developed to reduce greatly the chance of a hazardous situation occurring. It has been found that for the rat, the LD_{50} value for sugar is around 33 g/kg bodyweight (this drops to just 3.5 g/kg bodyweight for table salt). If this result is translated to a 50 kg human, then the quantity of sugar needed to kill half of the recipients would be 1650 g. Thus, on adding sugar to tea or coffee (probably 4 g per teaspoon) the acute risk to our health is very small. In this trivial example, the risk assessment shows that there is a safety factor between the applied and possibly harmful dose of over 400, and sugar would be considered safe.

For fragrance ingredients, before a risk assessment can be undertaken, the route of exposure has to be considered. Although LD_{50} values are available for some fragrance ingredients, it is obvious that fragrances are not intended for consumption. Inhalation is the most obvious pathway to examine, as fragrances are produced because of their odour. However, it is well known that the human nose is capable of detecting a vast range of materials at levels measured in parts per million and less. For example, if 0.2 ml of a fragrance is applied to the skin (a dab behind each ear), it is detectable to the human nose for several hours after application (to anyone close to the wearer). For a fine fragrance that contains 10% of the fragrance compound, this means that only 0.02 ml of active ingredients are applied to the skin. If we now say that the fragrance can only be detected within 1 m of the wearer, then 0.02 ml of fragrance diffuses into 8 m^3 of air, equivalent to a concentration of 2.5 p.p.m. (assuming 0.02 ml of fragrance weighs 2×10^4 μg). This assumes that all of the fragrance evaporates immediately. As the fragrance is detectable for some hours after application, the actual concentration in the air must be lower than this, and the concentration of any individual ingredient even lower. If any of these ingredients were toxic by inhalation at these levels, then they would probably be used as chemical warfare agents rather than fragrances.

As inhalation and ingestion under normal-use conditions are of little consequence for the risk assessment, the role of the fragrance upon the skin must be examined as a route of exposure. In this case two effects need to be considered. The first is whether the material irritates when applied to the skin and the second is whether the material can penetrate through the skin and affect the immune system and organs of the body.

Skin Irritation

Although unpleasant, skin irritation, which manifests itself as redness (erythema) and/or swelling (oedema) is not a major problem, as once the source of irritation is removed the effect diminishes. Most fragrance ingredients are classed as mild or moderate irritants when in an undiluted form only, so the low levels found in consumer products are unlikely to be a serious source of irritation.

Skin Sensitization

Whereas it used to be thought that the skin was an impermeable barrier to the outside world, it is now known that a large number of chemicals are absorbed to some degree, including those used in fragrances (Hotchkiss, 1994). The actual amount that penetrates the skin depends on how much is applied to which area of the body, the area that it is spread over, whether the skin is covered or not and whether the product is rinsed from the skin before absorption can take place. The nature of the product being used can also affect the level of absorption, as it has been shown that the alcohol used as a carrier in fine fragrances and aftershaves enhances the penetration of some fragrance ingredients. Not that penetration *per se* is harmful, as many angina sufferers will testify. The absorbtion of glyceryl trinitrate through the skin used to be used as a rapid way of relieving the pain of this complaint.

For some fragrance ingredients, penetration through the stratum corneum into the epidermis can elicit an immune response leading to allergic contact dermatitis or skin sensitization. Initially, exposure to a skin sensitizer has no effect, but repeated exposure can induce an allergy, which can then occur on contact with the material at levels below that needed to have an irritant effect. This allergy usually appears 1–2 days after contact and often becomes more severe over the next 2–3 days. The reaction does not diminish as an irritant reaction does, but may last several days or even weeks after exposure. In severe cases the sufferer not only needs to avoid the cause of the allergy, but also any products that contain a significant percentage of it.

For those materials that RIFM has shown to be sensitizers, IFRA has either applied a restriction or banned their use (Table 10.1). In some

Table 10.1 *Examples of fragrance ingredients restricted by IFRA*[a]

Ingredient	*Restriction*	*Reason*	*% allowed on skin*
Acetylated vetiver oil	S	Sensitization	U
Acetylethyltetramethyltetralin (AETT)	P	Neurotoxicity	0
5-Acetyl-1,2,3,3,6-hexamethylindan	R	Phototoxicity	2
Angelica root oil	R	Phototoxicity	0.78
Bergamot oil expressed	R	Phototoxicity	0.4
Bitter orange oil expressed	R	Phototoxicity	1.4

Continued

Table 10.1 *Continued*

Ingredient	*Restriction*	*Reason*	*% allowed on skin*
p-t-Butylphenol	P	Sensitization and depigmentation	0
Cinnamic alcohol	R	Sensitization	0.8
Cinnamic aldehyde	Q	Sensitization	U
Citral	Q	Sensitization	U
Costus root products	P	Sensitization	0
Cumin oil	R	Phototoxicity	0.4
Cyclamen alcohol	P[a]	Sensitization	0
Dihydrocoumarin	P	Sensitization	0
Farnesol	S	Sensitization	U
Fig leaf absolute	P	Phototoxicity and photosensitization	0
Grapefruit oil expressed	R	Photosensitization	4.0
trans-Hept-2-enal	P	Sensitization	0
Hexahydrocoumarin	P	Sensitization	0
trans-Hex-2-enal	R	Sensitization	0.002
Hydroxycitronellal	R	Sensitization	1.0
Isoeugenol	R	Sensitization	0.02
Lemon oil cold pressed	R	Phototoxicity	2.0
Lime oil cold pressed	R	Phototoxicity	0.7
Limonene	S	Sensitization	U
6- and 7-Methylcoumarins	P	Photosensitization	0
Methyloctine carbonate	R	Sensitization	0.01
Musk ambrette	P	Neurotoxicity and photosensitization	0
Nookatone	S	Sensitization	U
Oppoponax	R	Sensitization	0.60
Phenylacetaldehyde	Q	Sensitization	U
Pseudoionone	P[a]	Sensitization	0
Rue oil	R	Phototoxicity	0.78
Safrole, isosafrole and dihydrosafrole	P[b]	Chronic toxicity	0
Styrax	R	Sensitization	0.6

R = Restricted, a use limit for consumer products is applied to this material; P = prohibited, this material is banned as a fragrance ingredient; S = specification, there is a defined grade, botanical source or method of production for this material; Q = quenching, this material can only be used in conjunction with an agent that prevents sensitization (IFRA, 1973); U = unrestricted, there is no restriction on the use of this material as long as it meets the defined specification; [a] there are exemptions for this material when it occurs as an impurity in another product; [b] there are exeptions to this restriction for essential oils containing these ingredients.

cases, IFRA has ruled that only material of a set purity or botanical source can be used in fragrances. For example, IFRA recommends that crude gums of American and Asian styrax should not be used as fragrance ingredients. Only extracts or distillates (resinoids, absolutes and oils), prepared from exudations of *Liquidamber styracifua* L var macrophylla or *Liquidamber orientalis* Mills, can be used and should not exceed a level of 0.6% in a consumer product. For acetylated vetiver oils, IFRA has recommended that they only be used as fragrance ingredients if they are produced by a method which gives products free from allergens. Such acetylated vetiver oils can be prepared using acetic anhydride:

—without a catalyst and at a temperature not exceeding 120 °C;
—with ortho-phosphoric acid at room temperature;
—with sodium acetate in toluene at reflux temperature.

The first two products can be used in their crude form after the usual washing procedures, but may be further purified. In the last case, distillation is necessary to give a suitable product.

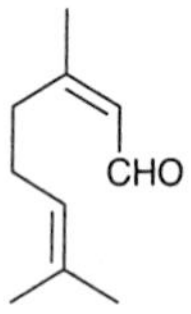

CHO
Citral
(5)

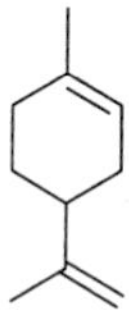

D-Limonene
(6)

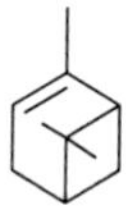

α-Pinene
(7)

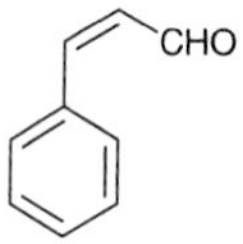

CHO
Cinnamic aldehyde
(8)

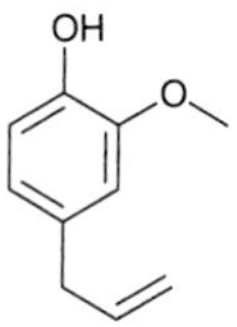

OH
O
Eugenol
(9)

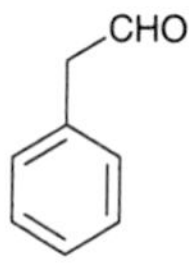

CHO
Phenylacetaldehyde
(10)

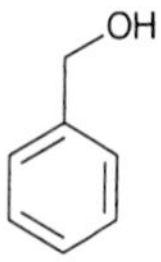

OH
2-Phenylethanol
(11)

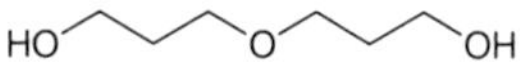

HO O OH
Dipropylene glycol
(12)

Table 10.1 gives a list of the materials that are banned or restricted by the guidelines. It is interesting to note that the sensitization caused by some fragrance ingredients can be suppressed. Neither lemongrass oil nor *litsea cubeba* oil are sensitizers although both contain over 70% of citral (5), which can cause sensitization at levels below 0.5% (Opdyke, 1976). Both of these oils also contain up to 20% of citrus terpenes, such as *d*-limonene (6) and α-pinene (7), which have been shown to prevent citral inducing a sensitization reaction (Opdyke, 1979), an effect which the industry refers to as 'quenching'. IFRA now recommends that citral be used only in conjunction with 25% of *d*-limonene or α-pinene. Other sensitizers that can be 'quenched' include cinnamic aldehyde (8) with an equal weight of *d*-limonene or eugenol (9) and phenylacetaldehyde (10) with equal weights of 2-phenylethanol (11) or dipropylene glycol (12).

Photoeffects

In some instances, it is not the material as applied to the skin that causes the allergic reaction, but a combination of the material and exposure to sunlight. In the late 1970s, the cause of an unusual number of cases of dermatitis in people using a sunscreen preparation was traced to the presence of 6-methylcoumarin, which was in the fragrance in the product (Kaidbey and Kligman, 1978). It has since been shown that a number of coumarin derivatives are capable of causing skin sensitization, but only when they are exposed to sunlight (see Table 10.2). As soon as it became apparent that 6- and 7-methylcoumarins and 7-methoxycoumarin were photosensitizers, their use was prohibited by IFRA.

Table 10.2 *Coumarin derivatives that are capable of skin sensitization when exposed to light*

Material	*Sensitizer*	*Photosensitizer*	*Photoirritant*
Coumarin (13)	No	No	No
6-Methylcoumarin (14)	No	Strong	No
7-Methylcoumarin (15)	No	Moderate	No
7-Methoxycoumarin (16)	Weak	Moderate	Strong
Dihydrocoumarin (17)	Strong	No	No
Hexahydrocoumarin (18)	Moderate	No	No
Octahydrocoumarin (19)	No	No	No

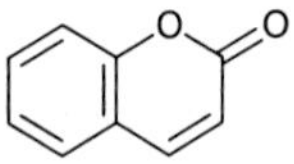

Coumarin
(13)

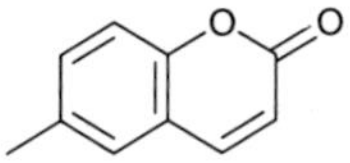

6-Methylcoumarin
(14)

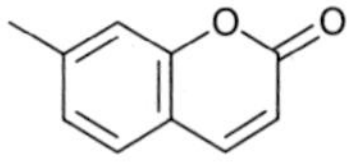

7-Methylcoumarin
(15)

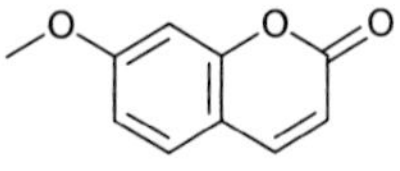

7-Methoxycoumarin
(16)

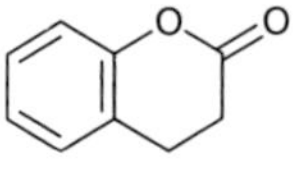

Dihydrocoumarin
(17)

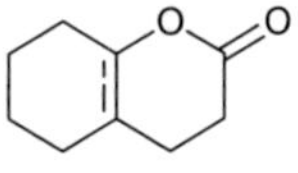

Hexahydrocoumarin
(18)

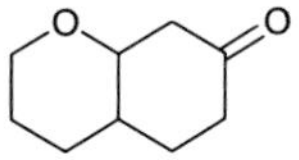

Ocahydrocoumarin
(19)

However, some essential oils, such as fig leaf absolute and certain citrus peel oils, contain natural phototoxic ingredients, usually referred to as furocoumarins (Marzulli and Maibach, 1970; Fisher and Trama, 1979). For example, some grades of Bergamot oil (expressed or cold-pressed varieties) are known to contain 5-methoxypsoralen (bergaptene, 20), which is a potent phototoxic agent (Zaynoun *et al.*, 1977). The use of these grades of oil, which can contain up to 0.3% of bergaptene, is restricted by IFRA, although the industry now tends to use citrus oils from which the furocoumarins have been removed by distillation or extraction. Fig leaf absolute is banned from use, for even at levels of 0.001% it can still elicit a photoallergic reaction (Opdyke, 1982).

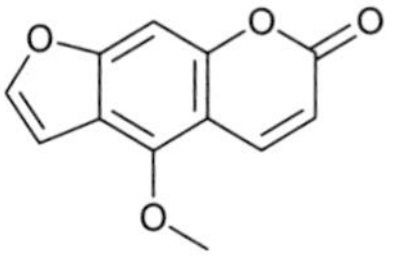

5-Methoxypsoralen
(20)

Neurotoxicity

It is obvious that if a fragrance ingredient can penetrate through the skin, it must have the potential to affect all the body systems, not just the immune system. This was found to be true for two materials that both exhibited neurotoxic effects. The first material, 6-acetyl-7-ethyl-1,1,4,4-tetramethyltetralin (AETT, 21), a synthetic musk, was found to cause damage to the nervous system of rats when it was applied dermally at a level of 9 mg/kg bodyweight per day, over a 26-week period (Spencer *et al.*, 1979; Ford, 1994). Although RIFM established a No Observable Adverse Effect Level (NOAEL), the expert panel decided that there was insufficient evidence to recommend the continued use of this material, and acting on this advice AETT was banned by IFRA. It is interesting that the related and very widely used musk material 7-acetyl-1,1,3,4,4,6-hexamethyltetralin (22) does not show any neurotoxic effects. When neurotoxic effects were reported for the second material, Musk Ambrette (23), RIFM undertook a research program to enable a comprehensive risk assessment to be made (Spencer *et al.*, 1984).

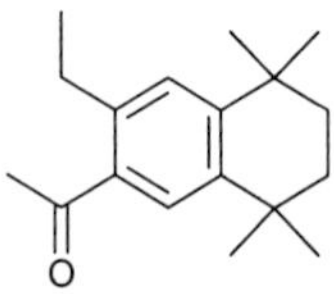

6-Acetyl-7-ethyl-1,1,4,4-tetramethyltetralin
(21)

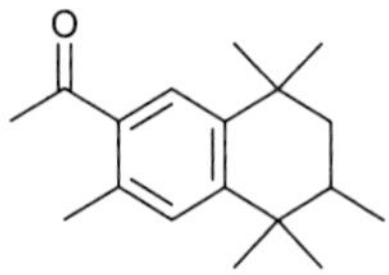

7-Acetyl-1,1,4,4,6-hexamethyltetralin
(22)

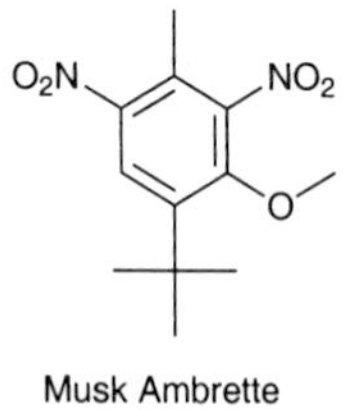

Musk Ambrette
(23)

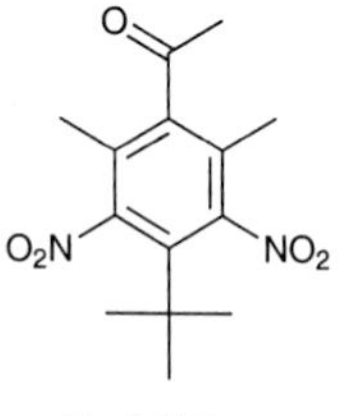

Musk Ketone
(24)

RIFM first established a dose level that a consumer may be exposed to by normal usage of a wide range of fragranced consumer products that contained musk ambrette. This level, which was weighted to the maximum likely exposure, was calculated to be 0.3 mg/kg bodyweight per day. RIFM also determined that musk ambrette had a NOAEL of 10 mg/kg bodyweight, which gave a safety factor of only 33 (NOAEL/daily exposure). However, skin penetration studies had shown that only 2% of an applied dose of musk ambrette was actually absorbed through the skin. Thus, the dermally applied daily exposure figure was really 50 times lower than calculated, giving a 50-fold higher safety factor. This evidence would almost certainly allow the industry to continue using musk ambrette, but when it was also reported to have photoallergic effects it was banned from use by IFRA (Cronin, 1984). It is interesting to note that other nitrated musk ingredients, such as Musk Ketone (24) and Musk Xylene (25) do not show any evidence of neurotoxic or phototoxic effects (Lovell and Sanders, 1988; Ford and Api, 1990).

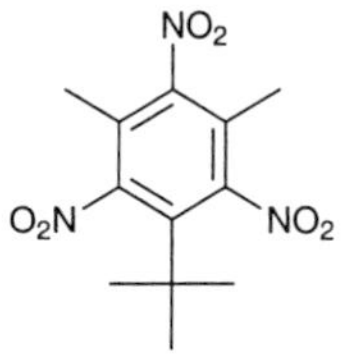

Musk Xylene
(25)

Reproductive Effects

Any chemical, including a fragrance ingredient, that enters the bloodstream could affect the reproductive system and/or the growing foetus, leading to a range of problems including infertility and birth defects. Thankfully, there has only ever been one case of a fragrance ingredient being implicated as causing adverse effects in the offspring of pregnant rats (Mankes *et al.*, 1983). The ingredient was phenylethanol (PEA, 11), which is the major component of rose oil and has been used in perfumery for several centuries. When RIFM examined the results of this study, in which PEA was fed in enormous quantities to pregnant rats by way of a stomach tube (gavage), it concluded that a more appropriate study on dermally applied PEA should be undertaken (Spencer *et al.*, 1979). The results of these studies, which included skin penetration and metabolism studies, showed that the PEA that did penetrate the skin was quickly metabolized to phenylacetic acid (PAA, 26), a natural component of human blood. It was concluded that the very tiny increase in PAA level in the blood stream from the use of PEA posed no significant risk to the consumer, and combining the results of other studies, a safety factor of many thousands was calculated. The adverse effects seen in the original study highlight the difficulties of extrapolating academic animal studies to the actual conditions encountered by the consumer.

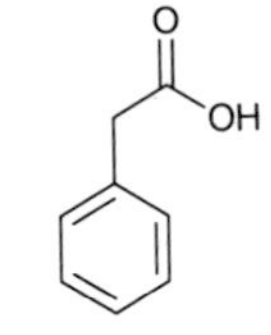

Phenylacetic acid
(26)

Natural Ingredients

From a recent trend in product labelling and advertising in which 'naturalness' is extolled as a virtue, it may be concluded that natural materials are safer than synthetic ingredients. This is obviously a false conclusion, as reference to Table 10.1 shows a number of natural materials that are banned or restricted by IFRA. Both hemlock and botulism are natural, but neither could be classed as safe for consumer use. Natural fragrances are no safer than those made with synthetic ingredients, and this 'naturalness' concept is not supported or promoted by the fragrance industry. It could well be argued that synthetic materials are safer than natural materials, since they are produced to very exact quality standards, whereas the quality of many essential oils varies greatly due to the vagaries of Mother Nature.

CONCLUSIONS

From the examples quoted above, it is clear that the use of fragrances poses very little risk to the consumer. Although there have been very few problems associated with fragrance use, RIFM and IFRA continue to examine all the available evidence on both traditional and new fragrance materials to ensure that the consumer, including those who buy 'Eve', can continue to purchase fragrances without having to worry about their safety.

REFERENCES

E. Cronin, *Contact Dermatitis*, 1984, **11**, 88–92.
J. F. Fisher and L. A. Trama, *J. Agric. Food. Chem.*, 1979, **27**, 1334.
R. A. Ford, in *Perfumery Practice and Principles*, R. R. Calkin and J. S. Jellinek (eds), John Wiley, 1994, 441.
R. A. Ford and A. M. Api, *Food Chem. Toxicol.*, 1990, **28**, 55–61.
E. Guberan and L. Raymond, *Br. J. Ind. Med.*, 1985, **42**, 240–245.
S. Hotchkiss, *New Scientist*, 1994, **1910**, 24.

IFRA 1973, *IFRA Code of Practice*, The British Fragrance Association, 6 Catherine Street, London, WC2B 5JJ.
K. H. Kaidbey and A. H. Kligman, *Contact Dermatitis*, 1978, **4**, 277.
W. W. Lovell and D. J. Sanders, *Int. J. Cosmet. Sci.*, 1988, **10**, 279.
R. Mankes, R. LeFevre, H. Bates and R. Abraham, *J. Toxicol. Environ. Health*, 1983, **12**, 235.
F. N. Marzulli and H. I. Maibach, *J. Soc. Cosmet. Chem.*, 1970, **21**, 695.
D. L. J. Opdyke, *Food Chem. Toxicol.*, 1976, **14**, 197.
D. L. J. Opdyke, *Food Chem. Toxicol.*, 1979, **17**, 259–266.
D. L. J. Opdyke, *Food Chem. Toxicol.*, 1982, **20**, 691.
P. S. Spencer, A. B. Sterman, D. Horoupian and M. Bischoff, *Neurotoxicology*, 1979, **1**, 221.
P. S. Spencer, M. C. Bischoff-Fenton, O. M. Moreno, D. L. Opdyke and R. A. Ford, *Toxic. Appl. Pharmac.*, 1984, **75**, 571.
S. T. Zaynoun, B. E. Johnson and W. Frain-Bell, *Br. J. Dermatol.*, 1977, **6**, 475.

Chapter 11

Volatility and Substantivity

KEITH D. PERRING

PERFUME CREATION AND PHYSICAL CHEMISTRY

The perception of a perfume depends, in the first place, upon the presence of odorant molecules in the air, and upon their nature and concentration. Most perfume starts off life as a liquid comprising a wide variety of molecules and of a known composition. In general, perfumers do not have a corresponding knowledge of the composition of volatiles in the air above such a mixture, except on those occasions where headspace analysis has provided hard analytical data (see Chapter 12). Perfumers therefore have to build up knowledge bases that summarize the olfactory behaviour of hundreds of ingredients under many different circumstances:

—in admixture with diverse permutations of other perfume materials (commonly 20–100 different feedstocks, any of which may themselves be complex oils);
—in product forms ranging from aqueous liquids to powders, creams, solids, *etc.*;
—on substrates such as hair, skin, cotton, *etc.*

Gaining this sort of experience is a lifetime's work for a perfumer, but it is possible to help the process by understanding the factors that govern the physical performance of perfume. One of the challenges for the physical chemist is to be able to predict what happens when a perfume is incorporated into different products; in particular, how does the composition of the perfume headspace alter, and what will be the 'in-use' perfume behaviour (for example, how to maximize a perfume's useful lifetime on skin, or ensure that a laundry perfume not only supplies a pleasant smell in the detergent, but also on the fabric after drying).

It is known that as soon as a perfume is incorporated into a product matrix, pronounced effects occur on some ingredients, those for which the 'microenvironmental' interactions have changed significantly. Physicochemical interaction is a consideration no matter what the base, whereas chemical interactions only become really important in consumer products possessing a reactive challenge for fragrance, for example products with low or high pH (*e.g.* below 4, above 9), or redox power (*e.g.* cold wave hair products or laundry bleaches). (Chemical stability issues are elaborated in Chapter 9.)

Physical fragrance–base interactions, which are very different to those experienced by fragrance ingredients when in the 'bottle', affect the profile of odorants in the headspace, and probably also have consequences for the evaporation characteristics. It is easy to see how complex this situation is, particularly in view of the ever-changing composition of the perfume as it evolves during evaporation. Nevertheless, a number of broad generalizations based on physical properties can aid the perfumer in the selection of ingredients for any particular product and desired performance criteria. Some of the physicochemical properties that underlie ingredient behaviour are highlighted below, and are part of the knowledge base the perfumer brings to bear. In the particular case of the Business Scents Ltd brief, this knowledge here can be added to by studying the fragranced products analytically to quantify actual perfume headspace, and by studying the delivery profiles of the various perfume components to the target substrates, skin and hair. In this way, a broader range of perfume ingredients may be identified that will perform better from the point of view of transport and distribution – after that, it's back to the perfumer.

PERFUME INGREDIENT VOLATILITY

The term 'volatility' is usually taken to refer to the speed at which a material evaporates. It is not an exactly defined property, and no universally accepted standards are laid down within the scientific literature. As implied above, the 'dry-down' or evaporation behaviour of even an unsophisticated perfume on a simple solid substrate, such as a paper smelling strip, is complex. Some materials are so volatile that they are lost much more rapidly than other components (within minutes), while materials at the opposite end of the volatility spectrum may remain for a considerable time (weeks or months). However, there is no doubt that when these very different types of molecule are present in the same mixture (as is the case for the majority of perfumes), ingredient interactions modify the evaporation behaviour to some extent, and, for those materials with intermediate molecular properties, the constitution of the perfume can govern their behaviour.

An understanding of the inherent tendency of an ingredient to escape into the gas phase is a useful starting point when considering perfume volatility. To a first approximation, the relative molecular mass (RMM) and the boiling point of a perfume ingredient provide some guidelines to behaviour. For materials for which boiling point data are not available, it is generally a sound alternative to look at chromatographic behaviour. For example, the retention time for a material to elute through a gas chromatographic column containing a non-polar phase is often strongly related to boiling point (in fact, such columns are commonly referred to as 'boiling point' columns).

Let us examine the relationship between boiling point and molecular size more closely. Table 11.1 comprises physicochemical information on a number of materials that are or have been used in the fragrance industry. The data were drawn from a number of sources, and some of the parameters (*e.g.* 'log *P*' and '*sp*', which are described later) were calculated from specific mathematical models, so that slightly different

Table 11.1 *Representative physical properties of perfume ingredients*

Ingredient	*RMM*	*Boiling point* (°C)	*Vapour pressure* (mmHg)	sp ($MPa^{0.5}$)	*log* P
Benzaldehyde	106.1	178	1.10	21.9	1.50
1,8-Cineole (3)	154.3	176	1.65	16.6	3.22
Cervolide® (2)	256.4	170 (5 mmHg)	N/A	19.0	4.54
Ethyl propanoate	102.1	99	36.5	18.2	1.21
Lilial®	204.2	127 (6 mmHg)	0.0045	18.5	4.22
Limonene (1)	136.2	178	1.40	16.5	4.46
Methyl naphthyl ketone	170.2	170 (α-isomer) (20 mmHg)	0.0014	24.0	3.00
Methyl butanoate	102.1	102	30.2	18.4	1.18
2-Phenylethanol	122.2	218	0.11	23.7	1.52

RMM = relative molecular mass; boiling point is at *ca.* 760 mmHg unless otherwise stated; log *P* = common logarithm of estimated octanol/water partition coefficient (Rekker, 1977); *sp* = Hildebrand solubility parameter as calculated according to Hoy (Barton, 1985); vapour pressure is at 25 °C; Lilial® = 2-methyl-3-(4′-*t*-butylphenyl)propanal; Cervolide® = 12-oxacyclohexadecanolide.

values may be found in the literature. However, in the context of volatility and substantivity, the emphasis is not on absolute values but rather on understanding and quantifying the differences between molecules.

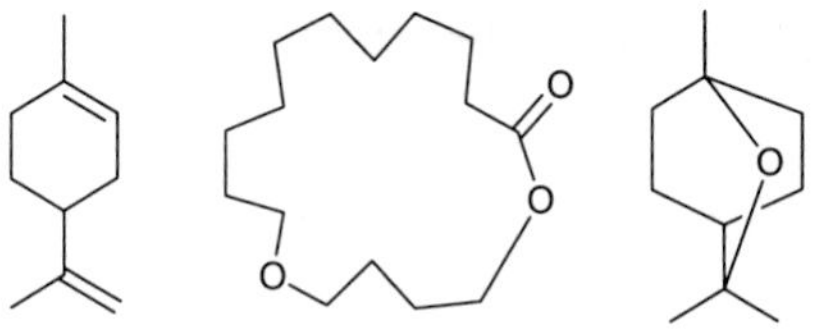

Limonene (1) 12-Oxacyclohexadecanolide (2) 1,8-Cineole (3)

The RMM of limonene (1), the major terpene in citrus oils, is 136, while its boiling point is *ca.* 178 °C at atmospheric pressure. Contrast this with Cervolide® (12-oxacyclohexadecanolide, a macrocyclic musk; 2), the RMM of which is 256 and boiling point is over 290 °C. The majority of perfume ingredients fall between these two extremes, although there are, of course, exceptions. Many odorous materials are more volatile than limonene and find some use in perfumes. Examples of these are methyl butanoate and its isomer ethyl propanoate, with boiling points of 102 °C and 99 °C respectively, and both with RMMs of 102. (These ingredients appear in a few perfumes, but are much more widely utilized in the flavour industry.) Odorous materials with RMMs greater than that of Cervolide® are much rarer since there appears to be a natural molecular size limit above which the human nose cannot detect, corresponding to around 300 RMM (*cf.* odorous steroids), presumably because volatility becomes too small.

Perhaps a more direct way to assess volatility is to look at the saturated vapour pressure of an ingredient. Saturated vapour pressure refers to the equilibrium pressure exerted by a substance in a closed system at a specified temperature (the volume of the system must, of course, be greater than that of the substance). Table 11.1 again gives representative values. Consider, for example, the volatile material 1,8-cineole (3), which is utilized in many 'fresh' perfumes and is also commonly found in toothpaste flavours. This material has a vapour pressure of *ca.* 2 mmHg at 25 °C (similar to that of limonene), which in the context of the perfumery world is very high. Most musks have vapour pressures that are three to five orders of magnitude smaller than that of cineole. Vapour pressure is directly related to the mass present in the gas phase, so the fact that musks are perceivable at all to the human olfactory system is a tribute to the impressive 'dynamic range' of olfaction (the sensitivity of canine olfactory systems is even better!).

Equation (1) is a useful route to calculating headspace concentrations above a pure substance from vapour pressures; c is the gas phase concentration in g l^{-1}, p^0 is the saturated vapour pressure in mmHg, and T is the temperature in Kelvin. Equation (2) is the same equation restated in terms of concentration m in mol l^{-1} at a temperature of 25 °C for cases where the molecular mass is unknown. These equations derive directly from the ideal gas equation.

$$c = 0.01604\,(p^0)\,(RMM)/(273.2 + T) \quad (1)$$

$$\log_{10}(m) = \log_{10}(p^0) - 4.269 \quad (2)$$

So far, we have dealt with pure materials. When liquid mixtures are considered, the headspace composition reflects the constitution of the liquid phase. Each component of the mixture is present in the gas phase, but its concentration depends on the nature and concentration of the other components. Clearly, when an ingredient is incorporated into a liquid mixture, the same amount is no longer present in the headspace (assuming that it is truly diluted, *i.e.* that the system is homogeneous with no phase-separated droplets). The saturated vapour pressure still gives a useful guide to the concentration in the headspace, as is evident from equation (3), where p is the partial vapour pressure of the ingredient, x its mole fraction in the liquid and p^0 its saturated vapour pressure. The parameter γ is known as an activity coefficient, and may be considered as an indicator of non-ideal behaviour.

$$p = \gamma x p^0 \quad (3)$$

When γ is unity, equation (3) reduces to the 'ideal' form known as Raoult's Law, in which the partial pressure of a component above a homogeneous liquid system is directly proportional to its mole fraction. It is very similar to an empirical expression first formulated by Henry for solutes in dilute solution (originally for gases in liquids), in which the solute partial pressure is proportional to concentration [equation (4), where c is the concentration in g l^{-1} and H is referred to as the Henry's Law constant].

$$p = Hc \quad (4)$$

By re-writing equation (4) in terms of mole fraction (and adjusting the dimensions of the constant H), the similarity to Raoult's Law can be seen. In fact, the main difference lies in the choice of standard states for the definition of γ or H, but this is beyond the scope of this chapter. More importantly, we need to understand how and why γ varies from unity. The differences are driven by the interactions that take place between the various components of the mixtures, discussed in the next section.

PERFUME POLARITY

Mass alone is not the only determinant of volatility. Thus, benzaldehyde has an RMM of 106, close to that of methyl butanoate, but its boiling point is substantially higher (by 76 °C). Other factors are evidently important, and it becomes useful to deal with the set of molecular properties which collectively contribute to what is known as 'polarity'. Explicit definitions of this term are rare. It relates particularly to the degree to which electronic charge is spread evenly through a molecule or whether certain locations have relatively high concentrations of positive or negative charge. As explained below, we use it here to represent the summation of physical interactions at the molecular level which may influence a molecule's free energy and other thermodynamic parameters (and hence also its availability and mobility within any given matrix).

A number of molecular interactions are feasible, but not all are relevant for the types of molecules found in perfumes. For example, ion–ion forces are unlikely to feature in any direct perfume interactions, since fragrance components rarely bear a charge (except for materials that are influenced by extremes of pH, *e.g.* Schiff's bases are protonated in strong acid, carboxylic acids are substantially anionic above pH 6). The principal physical interactions contributing to overall polarity that need to be considered are ion–dipole, dipole–dipole, dispersion forces ('London' forces) and hydrogen bonding. These may all play a part, depending upon molecular size, the presence of permanent dipoles and the type of functional group present.

As a general guideline, usually the affinity between a molecule and its microenvironment is higher when it is surrounded by molecules that are capable of expressing the same types of interaction, *e.g.* limonene dissolved in a hydrocarbon or citronellol dissolved in ethanol. Some specific examples are discussed later, but the effect may be taken as a more general form of the often expressed chemical adage that 'like dissolves like'. The activity coefficient, γ, introduced in equation (3) provides a useful way of assessing affinity. For any single volatile material, values of γ greater than unity imply partial pressures in excess of that predicted by Raoult's Law, indicating that more material is in the gas phase than would be expected, and that the material is effectively being 'pushed' out of the system. Conversely, values of γ below unity imply lower partial pressures, less material in the gas phase, slower evaporation and, perhaps, a concomitant increase in the persistence of the material in a system. As an aside, certain high-boiling materials are well known to promote the longevity of other materials in a perfume. For the physical chemist this phenomenon (based on negative deviations from Raoult's Law) is an example of what is termed 'fixation' within the fragrance world, but to the perfumer the term also conveys harmonious blending of notes throughout a perfume's in-use life.

Table 11.2 contains some data that exemplify the above comments. It cites values of the activity coefficient for an ester (benzyl acetate), an alcohol (heptan-2-ol) and a terpene (limonene) in different environments encompassing a range of polarities: water (highly polar), an aqueous surfactant (as used in shampoos) and a moderately polar solvent used in perfumery (diethyl phthalate, DEP). It can be seen that heptanol and limonene have values of γ in DEP which are similar and also greater than that of benzyl acetate, but in water limonene has a very large activity coefficient. Limonene is not able to participate in any strong, cohesive interactions with water molecules, and the high γ is a consequence of this. Heptanol, on the other hand, can participate in hydrogen bonding and exhibits much lower values of γ. Note that in the case of the shampoo, we may be dealing with 'apparent' activity coefficients since the degree of liquid phase homogeneity is not certain (owing to the presence of micelles and/or emulsion droplets).

To reinforce what this means in practice, Figures 11.1 and 11.2 depict

Table 11.2 *Activity coefficients[a] at 40 °C of perfume ingredients in various media*

Ingredient	*Water*[b]	*Aqueous SDS*[c]	*DEP*[d]
Benzyl acetate	1750	75	0.3
Heptan-2-ol	2770	57	2.2
Limonene	(>70 000)[e]	732	2.3

[a] Data measurements on an ingredient mixture (taken from Behan and Perring, 1987); [b] under high dilution conditions (2 p.p.m.); [c] at 0.05% w/w dilution (SDS = sodium dodecyl sulfate at 10% w/w in water); [d] at 0.05% w/w dilution (DEP = diethyl phthalate); [e] various values appear in the literature; the figure quoted here is a minimum.

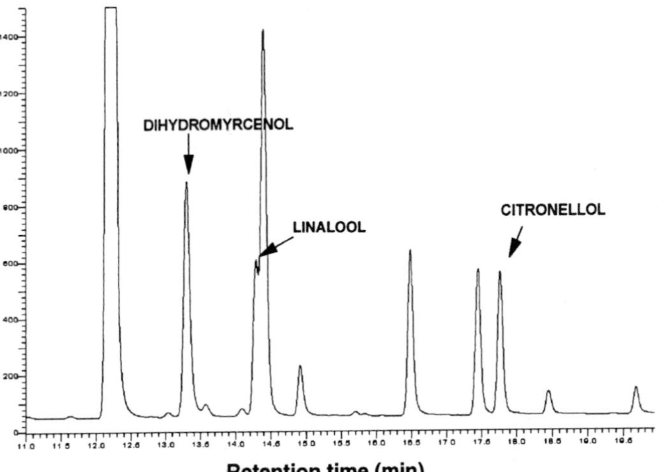

Figure 11.1 *Headspace chromatogram of perfume oil*

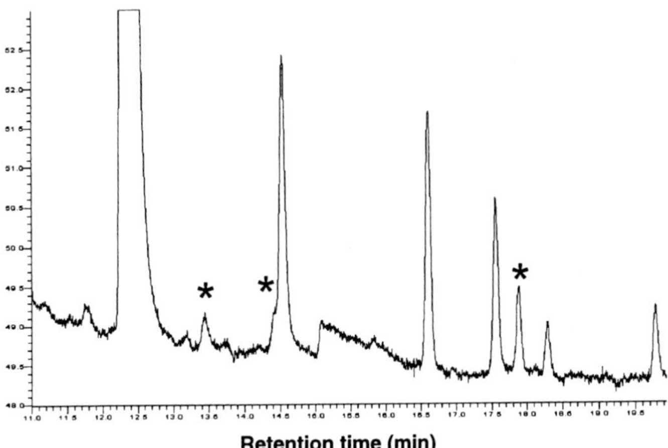

Figure 11.2 *Headspace chromatogram of cologne (2% of perfume in aqueous alcohol; asterisks denote peaks equivalent to labelled peaks in Figure 11.1)*

the equilibrium headspace profile (*i.e.* the gas phase concentrations of volatiles) above two systems containing the same perfume ingredients: a neat perfume oil and a cologne (typically these are alcoholic solutions containing 1–3% perfume). Figure 11.1 shows a (partial) headspace chromatogram containing labelled peaks corresponding to three terpene alcohols frequently used in cologne perfumes (dihydromyrcenol, linalool and citronellol). The same three peaks are marked by asterisks in Figure 11.2, which shows the headspace profile once the perfume oil has been taken up into aqueous alcohol. It can be seen that the concentrations of the terpene alcohols are relatively reduced in the polar medium (the cologne) compared with non-hydrogen-bonding molecules bearing other functional groups. Similar, but smaller, differences are also present for many of the other materials, and in consequence differences in odour characteristics may well occur. This level of understanding is useful, but the perfumer needs to know how to select ingredients that provide superior performance in the target product area. To meet these needs, we need to recognize and ideally measure or calculate the molecular characteristics that govern or mediate activity coefficient behaviour.

It would be very convenient if it were possible to calculate activity coefficients for any molecule in any given environment. Unfortunately, few situations occur in which this is possible, and rarely do such situations appertain to real products. Thus, for example, we may estimate the activity coefficients of many alkanes and simple derivatives at infinite dilution in water, but the corresponding values for the same materials in a specific shower gel are not readily calculable from first principles. However, a large number of parameters are available in the literature and have been used to answer questions related to physical behaviour.

The pharmaceutical industry has for many years developed mathematical models to explain biological activity of drugs: these are termed quantitative structure–activity relationships (QSARs). These techniques may also be applied to the situations described herein, although more correctly we are often more interested in QPARs, where the P stands for 'property' (which may refer to macroscopic properties, such as density, melting point, or viscosity, or to molecular or sub-molecular properties, such as molecular or fragmental volumes, or atomic charges). Since these properties are always at least partially dependent upon 'structure' in its broadest sense, the distinction between these approaches is sometimes blurred, particularly when pure structural data are used in the same mathematical model as a melting point! But whatever the semantics, the underlying need remains the same: to understand and predict the effect of changing the structure of a perfume ingredient on its performance, so that we may reasonably estimate, for example, the concentration of a perfume ingredient above a cologne, or its affinity for cotton during a wash cycle. This whole approach is detailed elsewhere in this book (see Chapter 14), so just two parameters found to have widespread application in the perfumery area are discussed here. These parameters are the octanol/water partition coefficient (usually expressed as its common logarithm, log P), and the Hildebrand solubility parameter (*sp*, often designated δ with or without a subscript). The Hildebrand parameter (Barton, 1985) is defined in equation (5), where ΔH is the molar enthalpy of vaporization, V is the molar volume, R is the gas constant and T is the temperature. The SI unit for the solubility parameter is $\text{MPa}^{0.5}$, but the c.g.s. system equivalent $(\text{cal cm}^{-3})^{0.5}$ is commonly seen.

$$sp = [(\Delta H - RT)/V]^{0.5} \tag{5}$$

The physical significance of these parameters is discussed in the next section. For now, the key point to make is that these parameters may feature explicitly in empirical QPARs developed to help ingredient selection or design. Alternatively, they may be used simply as classification variables to help identify ingredients that are likely to exhibit good performance in a specific fragranced product. In this second approach, a number of parameters are investigated for their usefulness in characterizing behaviour, *e.g.* a plot of log P versus boiling point for a variety of ingredients may lead to the identification of clusters of materials with good performance. Such classification variables may be used qualitatively or quantitatively, depending upon the difficulty of the problem and the statistical expertise available. The same parameters and approaches find particular use in the development of perfumes in which fragrance longevity is a key requirement, and this is the focus of the next section.

SUBSTANTIVITY AND RETENTION

We have touched briefly on how reduced vapour pressures can lead to reduced rates of evaporation. This means that physicochemical properties not only influence perfume volatility *per se* but also affect other aspects of fragrance behaviour, such as substantivity and retention. These terms are used within the industry to denote perfume longevity in use, usually with respect to a particular substrate and/or surface (*e.g.* skin, hair, cloth, *etc.*).

To understand perfume behaviour on these surfaces and/or matrices, we must consider the range of attractive or repulsive forces between the perfume components and the surface itself. The situation is complicated by the way in which perfume is delivered to the surface. For example, for a perfume ingredient in a soap bar to be substantive it must first be efficiently delivered to the skin during washing, it must then survive rinsing and, finally, it must be retained for some time on the skin. Definitions of 'substantivity' and of 'retention' vary, but here 'retention' is used to indicate the affinity a perfume has for a substrate when delivered to it, whilst 'substantivity' also includes delivery barriers.

As implied above, the delivery of at least some perfume to a surface is, except perhaps for air fresheners, a key requirement for perfume longevity in use. For certain products, such as cologne or deodorant, perfume delivery is very effective since the perfume is applied directly to the target substrate (the skin). Delivery from other products, such as soap or laundry powder, involves perfume transferring from an aqueous detergent solution or dispersion to a substrate or surface. This process has analogies with partitioning of materials between different phases, and, perhaps not surprisingly, we find that the octanol–water partition parameter (log *P*) may often provide insights into physical behaviour. For example, in dilute wash systems the deposition of ingredients onto substrates is often moderately or strongly correlated with log *P*. It is likely that the log *P* values reflect the solubilities of the perfume ingredients. By definition, materials with high log *P* such as the macrocyclic musks (log *P* values typically of 4.5 to 6.5) tend to be hydrophobic ('water hating') and partition where possible into more lipophilic ('oil loving') phases. Conversely, materials with low log *P* such as 2-phenylethanol (log *P* of 1.5), are hydrophilic and either soluble or sparingly soluble in water.

In fact, several well known models in the literature allow solubilities in water to be estimated reasonably well from a knowledge of log *P* (and according to the functional groups present). For example, Hansch *et al.* (1968) has published several linear free energy relationships (LFERs) between molal solubility and log *P* for various classes of monofunctional molecules. The correlation coefficients for the LFERs (a measure of 'goodness of fit') were in the range 0.93–0.99, indicating that solubility estimations, at least for some classes of material, are likely to be relatively accurate.

We may conclude from the above that values of log *P* appear to give some guidance to the tendency of perfume ingredients to move from aqueous systems to (presumably) less polar surfaces (skin, hair, *etc.*).

Analogous partition data are available for solvents other than octanol, *e.g.* olive oil, and these may be more pertinent in certain situations. However, log *P* values based on octanol/water partition are easily accessible for thousands of substances and, for most of the compounds found in perfumery, may be estimated reasonably well using one of the mathematical prediction models described in the literature [the two most well known are those due to Rekker (1977), and to Leo *et al.* (1971)]. Additionally, it is often true that for many materials partition coefficients determined in different solvent–water systems often correlate strongly with one another.

Unfortunately, the log *P* of an ingredient does not always suffice to describe behaviour adequately. For example, in concentrated aqueous detergent systems it is probable that perfume partitioning into complex surfactant phases becomes dominant, and knowledge of log *P* provides only a partial understanding. It then becomes necessary to search for other parameters which may be of more use. As mentioned above, a large number of parameters may be considered, but here we look more closely at just one, the solubility parameter (introduced earlier).

The Hildebrand solubility parameter has its origins in the development of what is known as 'regular solution theory'. As can be seen from equation (5), it is essentially a measure of how much energy is needed to disrupt intermolecular cohesion: the higher the *sp* value the more cohesive the material ('sticky' at the molecular level), and the harder it is to separate into individual molecules in the gas phase. Originally, *sp* was exploited primarily in the paint and polymer industry, but has since been found useful across a number of applications (Barton, 1985), too numerous to discuss here, but dealing with properties such as viscosity, surface adhesion, miscibility and, of course, volatility. As a working rule, different molecules with similar values of *sp* are likely to have significant interaction. This is similar to some of the conclusions made above when discussing polarity, and it also suffers from the same drawback, *viz.* the overall interaction may be complex, deriving from a superposition of mechanisms. It is possible to resolve *sp* into different components reflecting different interactions (*e.g.* hydrogen bonding, dispersion, *etc.*), and these may sometimes be more useful than the overall *sp* value.

The *sp* values of most perfumery ingredients fall between *ca.* $16\,MPa^{0.5}$ (non-polar materials such as terpene hydrocarbons) and *ca.* $25\,MPa^{0.5}$ (polar materials such as alcohols). In general, we expect materials to have lower activity coefficients in microenvironments characterized by similar values of *sp*. For instance, limonene (*sp* value of 16.5) is expected to be compatible with plastics such as polyethylene and polypropylene (*sp* range typically 16–18), and to exhibit good solubility and retention in these polymers. We anticipate that other ingredients, such as phenylethanol (*sp* value of 23.7), would have less desirable interactions with these polymers, for example promoting phase separation, crazing, stress cracking, *etc*. On the other hand, in partially hydrolysed polyvinyl acetate (*sp* range typically 22–24), the situation is reversed. The solubility parameter thus finds good practical use for understanding perfume interactions with plastic packaging, as well as for providing a basis for understanding affinities in general.

It is only rarely that we have explicit values of *sp* for a surface or substrate of interest, but this does not impede study and model building from the perfumery perspective. The approaches outlined earlier, together with appropriate parameters capable of 'capturing' the major types of interaction present in a system, may all be used to help build up a picture of the key features in the delivery of perfume to skin, fabric, *etc*. Once on the target site, the affinity between various perfume ingredients and the site may be quantified analytically, and investigated theoretically. It is important to recall, however, that the single most important property for prediction of substantivity remains, except where partitioning is extremely discriminating, the ingredient vapour pressure.

CONCLUSIONS

In summary, the volatility and headspace behaviour of perfume components is broadly comprehensible in terms of molecular interactions, both within products such as shampoos and colognes, and on or within substrates such as cloth or hair. However, the extreme complexity of the interactions, and the number of components invariably present, renders it difficult to predict *a priori* the headspace compositions in any given situation. Similar comments also apply to the related phenomena of ingredient or perfume fixation and substantivity. Nevertheless, it is possible to:

—quantify perfume behaviour analytically for any given product and in-use combination;
—analyse the data obtained from the first stage to identify ingredients that perform well, and to seek (empirical) mathematical models that explain the behaviour for a particular system.

The knowledge and understanding gained in this manner is part of the cycle of learning, perfume creation and performance evaluation that is a fundamental element of modern perfumery.

REFERENCES

A. F. M. Barton, *Handbook of Solubility Parameters and other Cohesion Parameters*, CRC Press, Boca Raton, FL, 1985.

J. M. Behan and K. D. Perring, Perfume Interactions with Sodium Dodecyl Sulphate Solutions, *Int. J. Cosmet. Sci.*, 1987, **9**, 261–268.

C. Hansch, J. Quinlan and G. Lawrence, The Linear Free-Energy Relationship between Partition Coefficients and the Aqueous Solubilities of Organic Liquids, *J. Org. Chem.*, 1968, **33**, 347–350.

A. Leo, C. Hansch and D. Elkins, Partition Coefficients and their Uses, *Chem. Rev.*, 1971, **71**, 525–616.

R. F. Rekker, *The Hydrophobic Fragmental Constant*, Elsevier, Oxford, 1977.

Chapter 12

Natural Product Analysis in the Fragrance Industry

ROBIN CLERY

INTRODUCTION

In this chapter we follow the analytical work that might be carried out on some natural products to answer a typical enquiry from creative perfumery. In the second section we explore the different analytical techniques used in the fragrance industry, concentrating on their application to natural product analysis and the way in which they are used to provide the creative perfumer with information. Many analytical techniques also fulfil other roles in the fragrance industry. These roles are mentioned when discussing the techniques individually.

NATURAL PRODUCT ANALYSIS

The analysis of natural products is a well-established part of the fragrance industry. Historically, essential oil analysis has provided the creative perfumer with the information required to reconstitute the characteristic odour of an oil at a lower cost and has led to the identification of the key components responsible for the odour. An essential oil may contain 300 or more components, and many plants and fragrant natural products have yet to be investigated in detail. Some of these materials may be a source of new odours. The traditional methods of essential oil analysis coupled with the ever-increasing sensitivity of modern equipment still leads to the discovery of fragrant natural chemicals that are new to science.

In Chapter 8, the creative perfumer approached the natural products analysis section for help in finding a natural lead for 'Eve', the new fragrance range for Business Scents Ltd. He is looking for an exciting, new, tropical, floral, fruity note to be included in this creation and has requested that some tropical fruit, muguet flowers and broom absolute be investigated. The enquiry can be answered using traditional essential oil analysis or by taking advantage of the recent developments in headspace analysis. The aim, in either case, is to identify a natural product that has the desired fragrance, and by extraction and analysis to provide the perfumer with the identity of its key odour ingredients.

The Traditional Approach

The traditional method of essential oil analysis is to extract the plant material by steam distillation or with solvent and then fractionally distil the oil or extract and isolate individual components by chromatographic techniques for subsequent identification by spectroscopic methods. At each step the odour of the fractions and isolates is assessed and those with the desired characteristics are investigated further. To answer the enquiry about the key odour components of broom absolute, first a sample of the absolute that is of an acceptable odour quality is obtained. The absolute is the alcoholic extract of the concrete, which is itself the solvent extract of the flowers of *Spartium junceum*, Spanish broom, often referred to by its French name Genêt. The odour of any natural extract can vary according to the geographical origin and quality of the plant material, the time of year it is harvested and the extraction method used. If no sample of adequate quality is commercially available then the fresh flowers would be obtained from the plant and the extraction carried out in the laboratory.

The first step towards separating the components of the absolute is fractional distillation. The absolute is distilled in the laboratory under vacuum, with a nitrogen blanket to reduce the risk of thermal degradation of susceptible compounds. The individual fractions collected from the distillation are analysed by gas chromatography–mass spectrometry (GC–MS) to determine their composition; some fractions might contain a single component while others may still be complex mixtures. Each fraction is assessed by a perfumer to determine which retained the odour most characteristic of the original material. The chosen fractions are analysed by GC-sniffing. If any of the individual components of the mixture are attributed with the characteristic odour that the perfumer requires, then work is directed at isolating and identifying these materials. If, however, the fraction is still too complex it is further fractionated by some form of liquid chromatography, such as flash chromatography, and the sub-fractions are assessed by a perfumer and analysed by GC–MS and GC-sniffing.

If materials are found that have an interesting odour, but that cannot be identified by GC–MS, they are isolated by preparative techniques such as preparative high-performance liquid chromatography (HPLC) or preparative GC and their structures determined by nuclear magnetic resonance (NMR) which, in most cases, is able to provide an adequate identification. As the compositions of the chosen fractions and sub-fractions are determined, the perfumer tries to create an accord which adequately represents the odour of the original material by reconstructing the fractions from their individual components.

The Headspace Approach

To answer the enquiry about the fragrance of muguet (lily of the valley), it would be most appropriate to use headspace analysis. Some natural materials, especially flowers and fruits, are often not available in sufficient quantity for even a laboratory extraction, while others yield an extract that does not reflect the fragrance of the flower. It is for these flowers that headspace analysis has great advantages. Using the non-destructive headspace trapping technique and light, portable sampling equipment, the fragrance of any flower, fruit or other natural source can

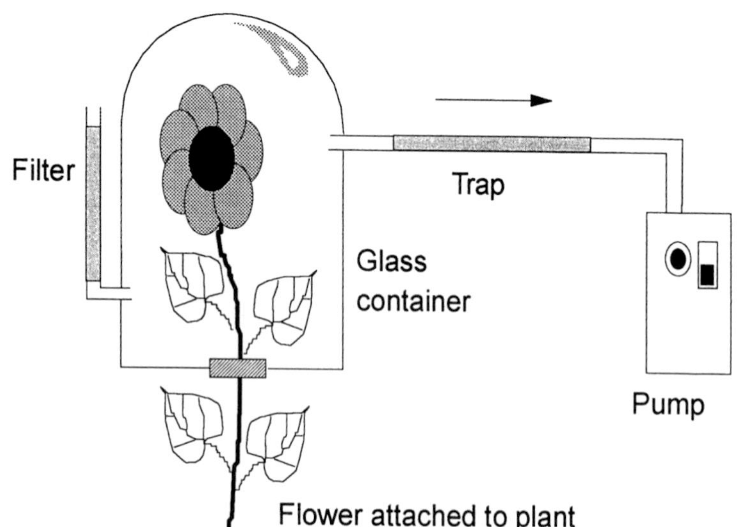

Figure 12.1 *Portable headspace sampling equipment*

be collected from the field (or greenhouse) and returned to the laboratory for analysis without disturbing the flower or plant (Figure 12.1).

Muguet is a popular garden flower, so finding plants could be as easy as visiting a garden centre or asking friends. Once some suitable flowers have been found, the headspace equipment is taken to the garden, the special glass bell jar carefully set up over the flowers and the headspace collected on several different types of traps. First, about four small thermal desorption traps are used, in which enough sample can be collected in a few minutes, then some larger-capacity solvent desorption traps are left in place for about 8 hours each. This ensures that a representative sample of the odour is collected and provides appropriate samples for the different analytical techniques to be used.

Back in the laboratory, the thermal desorption traps are analysed by GC–MS and GC–FID (flame ionization detector), by which means the major components can be quantified and identified. This reveals that the major components of the headspace of muguet are benzyl alcohol, citronellol and citronellyl acetate, and that the minor components with important odours are myrcene, *trans*-ocimene, *trans*-rose oxide, and 6-methylhept-5-en-2-one. Usually, many compounds in the sample cannot be identified definitively by GC–MS. Therefore, the larger traps are desorbed with diethyl ether or a similar solvent, and the resulting solution analysed by GC-sniffing to determine which components had a muguet odour. The solution is also concentrated and analysed by GC–MS to investigate the trace components. The use of nitrogen- and sulfur-specific detectors facilitates the identification of nitrogen- or sulfur-containing compounds, such as indole (1) and benzyl cyanide (2), which have, indeed, been found in muguet.

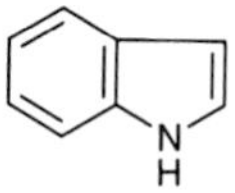

Indole
(1)

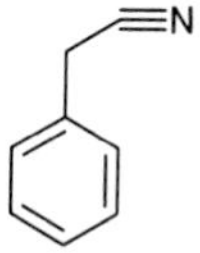

Benzyl cyanide
(phenylacetonitrile)
(2)

The results passed on to the perfumer would include the name of the compound, its relative proportion in the headspace sample and an indication of the odour intensity of the material. The perfumer can then select the important odour materials and combine them in an accord to re-create the scent of the flower.

To complete an analysis of fresh, tropical fruit, we took advantage of the Quest office in Indonesia and asked an employee who was travelling to Indonesia to take the equipment, since it is easily portable. With the help of the local perfumers the equipment was set up to capture the fragrance of the best smelling fruit in a local market. The traps containing the samples were sent back to the laboratory, where the analysis revealed exactly which compounds contributed to the real freshness of the fruit. The GC–MS and GC-sniffing results showed that methyl 3-methylbutanoate, methyl 3-methylpentanoate, methyl 3-methylpent-2-enoate, and methyl *trans*-hex-2-enoate are found in these fruits and contribute significantly to their odour. By providing the results of these analyses to the perfumers, we are able to help them understand the construction of the natural fragrances and re-create the original odour more closely.

Modern natural product analysis reveals both the chemical composition of new oils or flower scents and the identity of novel fragrant molecules that may become new perfumery ingredients in the future. It is chemical detective work to solve the mysteries of nature's fragrances that have evolved over the millennia.

ANALYTICAL TECHNIQUES USED IN THE FRAGRANCE INDUSTRY

Extraction

Steam distillation is the main commercial extraction procedure for the production of essential oils from almost any type of plant material. Solvent extraction is also used commercially and yields a resinoid, concrete or absolute according to the solvents and techniques used (see Chapter 4). Both steam distillation and solvent extraction are used on a laboratory scale to produce oils and extracts for analysis. Other methods of extraction, such as supercritical fluid extraction (SFE), which uses supercritical CO_2 as the extraction solvent, are now being developed and used on both commercial and laboratory scales. The extracts produced by SFE may contain different materials from the steam-distilled oil because of the solvating power of CO_2 and the lower extraction temperature, which reduces thermal degradation. The CO_2 extract may therefore have an odour closer to that of the original material and may contain different fragrant compounds. The choice of extraction procedure depends on the nature and amount of material available, and the qualities desired in the extract. Solvent extraction is better suited to small sample amounts or volatile materials that could be lost during a distillation. Steam distillation is used in the laboratory if it is important to replicate the commercial process or to release all the volatile oils from deep within seeds, dried fruits or woody materials.

Gas–Liquid Chromatography

Usually referred to simply as GC, this is one of the most important and widely used analytical techniques in the fragrance industry, as it is ideally suited to the volatile compounds that are the basis of the industry. It is a means of separating a complex mixture into its components and subsequently quantifying the individual components. As a chromatographic technique, it is based on the partition of analytes between the mobile phase (a gas such as helium, hydrogen or nitrogen) and the liquid stationary phase, which is often coated onto the inner wall of a very narrow fused silica column. The time taken for a material to elute from the column (its retention time) is reproducible under identical operating conditions and, as the basic physical principles of GC are well-established (Ettre and Hinshaw, 1993; Hinshaw and Ettre, 1993) the behaviour of analytes can be predicted for different operating conditions. The retention time of a compound can be a good guide to its identity by reference to known standards.

The basic design of the gas chromatograph can be fitted with a range of specific injectors, columns and detectors to optimize the separation of components and aid their identification. Recent developments in computer control, the use of robotic autosamplers and the trend to couple instruments together for sequential procedures have lead to increased automation for routine analytical tasks performed by GCs in research, factory and quality control environments.

GC Injection Systems

Many specialized injection systems are used with the GC in the modern laboratory to deliver the complete sample to the column without alteration. The conventional injection method is simply to inject a small volume (about 1 μl) of a dilution of the sample using a syringe with a fine needle, which pierces a silicon rubber septum and delivers the sample into a heated chamber where it is vaporized and carried on to the column in the stream of carrier gas. Some of the carrier gas containing vaporized sample may be split from the main column flow and vented to reduce the amount of sample delivered to the column. This is known as a 'split injection' and is suitable for almost all liquid or soluble materials, even for samples that may contain trace amounts of non-volatile contaminants, as these are deposited in the vaporization chamber of the injector and do not reach the column.

The high temperature of the injector, typically 250 °C, means that this method is not suitable for analytes that are subject to thermal degradation. For these materials, an 'on-column' method is preferable, in which the solution of sample is injected directly into the narrow capillary column with a fine needle. On-column injection techniques are also more suitable for extremely dilute samples, as more sample is delivered to the column, but are less suitable for dirty samples containing non-volatile contaminants, which accumulate on the column.

One of the most versatile injection systems is the programmable temperature vaporizer (PTV), a design that incorporates a vaporization chamber with a rapid and accurately controlled heating element, such that the temperature can be programmed to rise following the injection; thus, analytes are swept onto the column without being subjected to temperatures above their boiling point. By electronically controlling the action of the split vent valve, the system can also be used to vent the volatile solvent from dilute samples at a low temperature and then to deliver the analytes to the column by increasing the temperature. This allows much larger volumes of dilute solutions to be injected, thus increasing the absolute amount of analyte on the column without flooding it with solvent.

The use of a headspace injection technique may be preferable if the analytes are contained within a non-volatile or corrosive matrix that cannot be injected directly (perfume in a washing powder, for example). The basic principle of headspace injection is the delivery of a volume of vapour from the space above the sample material to the GC column. This can be achieved in several ways:

—A gas-tight syringe.
—A sample loop and a system of valves to fill the loop with vapour and then direct it onto the column.
—An adsorbent on which the volatile materials are trapped, and subsequently desorbed with a solvent or by heating the adsorbent.

Only materials volatile at the sampling temperature are transferred to the GC. Therefore, most headspace injection systems include some means of either gently heating the vial containing the sample or bubbling a gas through the non-volatile liquid to purge the volatiles, which can then be trapped on an adsorbent. The purge-and-trap method of headspace injection is widely used for analysing very low levels of volatiles in water, whether they are environmental samples or samples of water containing perfume from washing machines or dishwashers. The trapping of natural volatiles on absorbent cartridges is the basis for headspace analysis of flowers and other natural materials (Ter Heide, 1985).

GC Columns

The most common form of GC columns used in the fragrance industry is the wall-coated, open tubular (WCOT) type. These are made of fused silica tubing coated with a thin film of stationary phase on the inside and covered with polyamide on the outside for protection. Fused silica columns are usually between 25 m and 60 m long, with inner diameters from 0.53 mm to 0.1 mm. A widely used range of stationary phases is based on phenyl-substituted methyl silicones. The polarity of the phase is determined by the number of phenyl groups, a greater proportion of which increases the polarity of the phase. OV-1 is a common 'non-polar' stationary phase composed of dimethyl siloxane, while SE-54 is a more polar phase in which 5% of the dimethyl groups have been substituted with phenyl groups (Figure 12.2). Phases with greater polarity are achieved by using polyethylene glycol (PEG). The range of GC stationary phases is being modified continuously and expanded

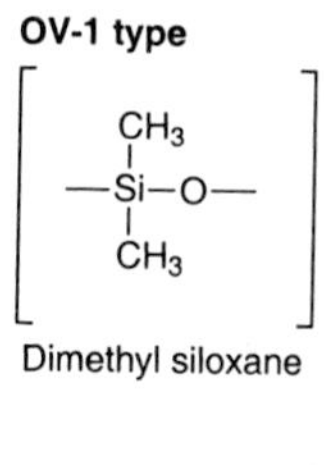

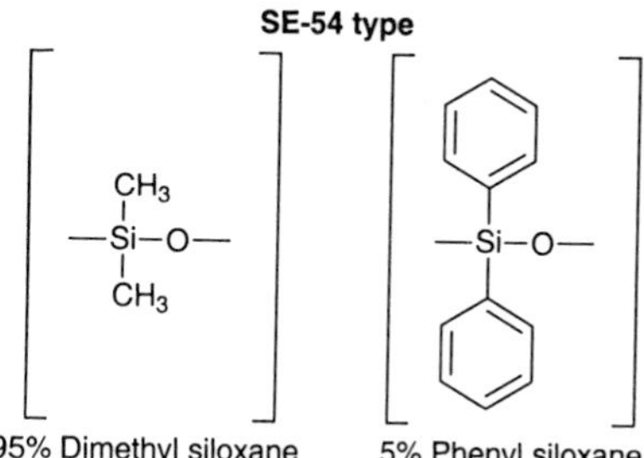

Figure 12.2 *Common GC stationary phases*

with new materials for specific applications. Cyclodextrin-based stationary phases are now available commercially for the separation of optical isomers, a technique referred to as Chiral-GC.

GC Detectors

The most common detector is the flame ionization detector (FID). The outlet from the column is directed into a carefully controlled hydrogen flame. The electrical potential across the flame is measured continuously, as this changes when an organic compound elutes from the column and is ionized in the flame. The analogue signal from the detector appears as a series of peaks over the duration of the analysis; these correspond to each material that elutes from the column. The amplitude of the signal is directly proportional to the amount of material passing through the detector and therefore the area under a peak on the recorded chromatogram is proportional to the amount of that material in the sample. The data are usually processed by an integrator or computer and the results are commonly expressed as a percentage of the total relative peak area (%RPA), the simplest quantitative measurement of the components in the mixture.

Coupling the column from the GC to a mass spectrometer provides a very powerful combination, GC–MS, which can identify and quantify almost all the compounds in a complex mixture, such as an essential oil or perfume, by reference to libraries of mass spectra of known compounds. Careful investigation of the mass spectrum can be used deductively to determine a possible structure for an unknown material using fragmentation theories to identify sub-structural components of the molecule. Recent developments in benchtop mass spectrometers have brought a range of specialized MS techniques into the realm of GC–MS machines; techniques such as chemical ionization and MS–MS are now available, which provide more information on individual sample components and allow better identification of unknown compounds.

GC-Fourier transform infra-red (GC-FTIR) spectroscopy is less frequently used than GC–MS, but involves a similar principle in which the outlet from the column is coupled to an infra-red spectrophotometer. The technique currently suffers from a lack of library spectra, as the IR spectra taken in the vapour phase can be subtly different from condensed-phase spectra or spectra collected using the well-established KBr disc method.

The use of nitrogen- or sulfur-specific detectors for GC enables small quantities of nitrogen- or sulfur-containing molecules to be detected.

These often have very powerful odours (Boelens and Gemert, 1994); for example, 2-isobutyl-3-methoxypyrazine from galbanum oil has a powerful green note and 1-*para*-menthene-8-thiol, a very powerful natural material, has a strong grapefruit odour at concentrations in the p.p.b. range.

GC-sniffing is an adaptation of special importance in the fragrance industry. The effluent from the column is split between a conventional detector and a smelling port that allows the individual components to be smelled by the human nose which is more sensitive to certain materials than sophisticated detectors (Acree and Barnard, 1994). If the nose belongs to a perfumer, then the odours can be recognized and described immediately. This is particularly useful when trying to establish the odour of a single component in a complex mixture, as GC-sniffing provides information from a few micrograms of sample which would only otherwise be available if the individual material could be isolated and purified in quantities of several grams (a very time-consuming process requiring relatively large amounts of sample).

Some separations can only be achieved by GC and if it is necessary to isolate such a material, then preparative GC is required. The flow from the column is momentarily directed to a cold trap as the desired compound elutes, which then condenses in the trap. The amounts that can be collected in this way are minute, but a few hundred micrograms are sufficient for a ^{1}H NMR or IR analysis.

Preparative Liquid Chromatography

Liquid chromatography (Hostettman *et al.*, 1986) in its many forms is a separation technique based on the polarity of the analytes and their partition between the mobile and stationary phases, and is therefore complementary to fractional distillation, which separates materials by their boiling point. The usual sequence for fractionating an essential oil or extract is to distil it first and then apply liquid chromatography to the distillation fractions as a further fractionation procedure, rather than as an analytical tool. The selectivity of the technique is achieved by choosing a stationary phase, usually from the various activities of silica gel, and varying the polarity of the mobile phase, the solvent, by mixing a non-polar component (such as hexane or pentane) with different amounts of a more polar component (such as diethyl ether, ethyl acetate or chloroform).

In the simplest form of liquid chromatography, the solvent is applied to the top of the column by gravity from a reservoir, a slow but reproducible process. Greater speed and resolution can be achieved, even with relatively large columns containing up to a few kilograms of silica, by using 'flash chromatography', in which air or nitrogen under pressure is used to force the solvent through the column. This is a convenient and reasonably reproducible method for quickly separating fractions of very different polarities. For greater resolution and reproducibility on a smaller scale, the solvent can be pumped through the column at a continuous and controlled rate, a technique variously called medium-pressure chromatography or preparative high-performance chromatography (Prep HPLC; Verzele and Dewaele, 1986), according to the equipment used. Detectors are not used with gravity or flash systems, but a non-destructive detector, such as a UV spectrophotometric detector or a refractive index detector (relative refractometer), is used with Prep HPLC to monitor the solvent stream and allow accurate collection of fractions as they elute from the column.

Solvent gradients can be used to improve resolution. A simple, stepwise gradient involves sequentially using solvents of increasing polarity and can be applied to any type of column chromatography. Continuous solvent gradients can be generated by modern HPLC pumps which mix solvents of different polarities to increase gradually the polarity of the mobile phase. Computer-controlled systems equipped with an autosampler and automatic fraction collector are available and can be programmed to repeat a separation many times and bulk the fractions from each separation. This allows larger amounts of sample to be processed, while achieving the resolution of a small-scale separation.

Analytical HPLC necessarily includes a detector on the outlet from the column, which responds to the presence of analytes in the solvent stream. Narrow-bore columns and fine particle sizes are used to achieve the best possible resolution. Although analytical HPLC is used in the fragrance industry to investigate the non-volatile fractions of essential oils and product bases, for example, it is much more widely used in other industries (such as the pharmaceutical industry, for which it is the main research tool).

POSITIVE IDENTIFICATION

Once a fraction or component of an essential oil or extract has been isolated, its identity needs to be determined. A familiar or common chemical is usually considered to be positively identified if its retention time on a given GC phase and its mass spectrum match those of a reference material. However, the identification of some materials with similar mass spectra, such as the sesquiterpenoids, requires retention times on two GC phases of different polarity. If the mass spectrum of a chemical is not in the reference libraries available and its structure cannot be conclusively deduced from its mass spectrum, then it must be purified and analysed by other spectroscopic techniques, such as NMR, IR or UV spectroscopy.

The advantage of NMR over MS is that the spectra produced are usually fully interpretable. A number of different experiments can be carried out using ^{1}H NMR and ^{13}C NMR, which normally generate enough data on the relationship of atoms within the molecule for the molecular structure to be elucidated. However, the exact structure and absolute configuration of a new natural material can only be determined by either complete synthesis or X-ray crystallography. Total synthesis of the material for comparison with the isolate is very time-consuming and may be very complicated for some complex natural materials and X-ray crystallography is an expensive technique only applicable to crystalline solids. NMR is, therefore, the most revealing analytical technique available for fragrance materials.

HEADSPACE COLLECTION

The headspace is the air above or around a fragrant substance that contains the volatile compounds. This can be collected for analysis when extraction of the volatiles from the material is not viable. This technique has been extensively developed for the collection and analysis of flower volatiles, since many flowers do not yield an extract that reflects the odour of the fresh flower while others are simply too rare to be available in sufficient quantity for extraction. Many different techniques have been applied to the collection of volatiles from the air above flowers, including the use of cold traps, solvent traps, adsorbent materials and capillaries coated with adsorbents (Ter Heide, 1985; Kaiser, 1991). However, the most common method for trapping flower volatiles is the use of adsorbent traps. The traps are small glass tubes containing activated charcoal, Tenax®, or a similar porous polymer through which the air is pumped. The volatiles from the air are adsorbed onto the trap while air and water pass through unretained, thus permitting the volatiles from many litres of air to be concentrated on the adsorbent. The flower is enclosed in a modified bell-jar or flask to prevent the volatiles being swept away by air currents and to isolate the flower from any contaminants that may otherwise drift into the sampling area. The traps are inserted into the enclosed space of the bell-jar and attached to a small, portable pump until sampling is complete.

The traps are then removed, sealed and returned to the laboratory for analysis. The duration and flow rate used for sampling are adjusted to match the expected concentration of volatiles in the air, the capacity of the traps and the intended method of desorption. Traps designed for thermal desorption contain only a few milligrams of adsorbent and sampling can be completed in a few minutes. In contrast, solvent desorption can be carried out on traps containing any amount of adsorbent from a few hundred milligrams to several grams. The traps have a finite capacity for individual compounds, dependent on the amount of absorbent and the dimensions of the trap. Different materials are retained to varying degrees according to their volatility and polarity, and it is therefore important to determine the capacity of the traps to ensure that none of the volatiles are lost during sampling.

Having collected the sample on the adsorbent traps it is important to keep them cool and dark, as the glass and absorbent surfaces are sufficiently reactive to cause degradation of some materials if exposed to light or heat for long periods. However, if adequately protected the samples can be stored for long periods or transported over long distances.

Samples collected in this way can be desorbed using a solvent which provides a solution for conventional injection on any GC or GC–MS. Thermal desorption of specially designed traps directly onto a GC with the appropriate injection system is more sensitive, and very volatile materials are not obscured by large solvent peaks. However, there is the possibility of thermal degradation of the sample and the entire sample is used at once.

The great advantage of headspace sampling is that it can be carried out relatively easily in the field with simple, portable equipment; this has opened up many exciting possibilities for the natural product chemist who can now analyse the scent of just a single flower. This technique has been successfully applied to a great range of plants and flowers for fragrance analysis, pollination studies and other botanical investigations (Kaiser, 1993; Knudsen *et al.*, 1993; Bicchi and Joulain, 1990). The sample size can be limiting when it comes to identifying a new material. Although modern instruments are extremely sensitive and only a few nanograms are required for acquisition of a mass spectrum, several hundred micrograms are required for ^{13}C NMR and isolating materials on this scale, even using GC-trapping, requires more sample than is usually available by headspace collection.

With the array of extremely sensitive instruments now available, the analytical chemist can identify materials of odour interest that are present at very low levels in natural materials. However, we still rely on traditional techniques and laboratory skills to fractionate samples and isolate new materials for identification.

REFERENCES

T. E. Acree and J. Barnard, Gas chromatography–olfactometry and charm analysis, *Dev. Food Sci.*, 1994, **35**, 211–220.

C. Bicchi and D. Joulain, Review: Headspace–gas chromatographic analysis of medicinal and aromatic plants and flowers, *Flav. Frag. J.*, 1990, **5**, 131–145.

M. H. Boelens and L. J. Van Gemert, Characterization and sensory properties of volatile nitrogen compounds from natural isolates, *Perfumer Flavorist*, 1994, **19**(5), pp. 51–52, 54–55, 58, 60, 62–65.

L. Ettre and J. Hinshaw, *Basic Relationships of Gas Chromatography*, Advanstar Communications, 1993.

J. Hinshaw and L. Ettre, *Open Tubular Column Gas Chromatography*, Advanstar Communications, 1993.

K. Hostettman, M. Hostettman and A. Marston, *Preparative Chromatography Techniques*, Springer-Verlag, 1986.

R. Kaiser, Trapping, investigation and reconstruction of flower scents, in *Perfumes, Art, Science and Technology*, P. M. Muller and D. Lamparsky (eds), Elsevier, 1991, 213.

R. Kaiser, *The Scent of Orchids*, Elsevier, 1993.

J. T. Knudsen, L. Tollsten and G. Bergstrom, Review article no 76: floral scents – a checklist of volatile compounds isolated by headspace techniques, *Phytochemistry*, 1993, **33**(2), 253–280.

R. Ter Heide, in *Essential Oils and Aromatic Plants*, A. Baerheim Svendsen and J. J. C. Scheffer (eds), 1985, 43.

M. Verzele and C. Dewaele, *Prep HPLC: A Practical Guideline*, TEC, 1986.

Chapter 13

Chemoreception

CHARLES SELL

BACKGROUND

The five senses (sight, hearing, taste, smell and touch) developed in living organisms as a means of giving them information about their environment. Chemoreception was probably the first of the senses to appear and is present in very simple, primitive species. The basic mechanism for detection of chemical signals from outside these primitive organisms is probably the basis of the senses of smell and taste which we and other higher animals possess today. Nature does not discard systems that work, but develops, adapts and refines them. This might give us clues about the receptor mechanism, since it probably began to evolve as a means by which aquatic organisms could detect water soluble chemicals. For most animals, the chemical senses are the ones on which they rely most heavily; it is only some groups of birds and a few primates that depend more on sight than on smell and taste. Despite the importance of the chemical senses, we understand very little about them. It is surprising that relatively little research into the chemical senses has been carried out, but encouragingly this is changing and many groups around the world are now beginning to take up the challenge and are already making significant headway.

The receptors used to detect odorants are similar to those used to detect hormones and, indeed, to those used in vision. Therefore, any increase in our knowledge of smell and taste receptors could also benefit our understanding in other fields, even though this is more advanced than our understanding of olfaction at present. Apart from the academic interest, there are obvious commercial reasons for increasing our understanding. For example, it is of particular interest is to be able to predict the odour of new molecules without having to prepare and evaluate them, and thus save time and money in the search for new perfume ingredients.

Perfumes were originally made entirely from natural chemicals; the essential oils, absolutes and concretes that could be obtained by extraction from vegetable or animal sources. With the introduction of organic synthesis in the nineteenth century, a variety of odorants became available as products of the chemical industry. Many of the early synthetic fragrance materials were the products of serendipitous discovery. For example, it was during his work on explosives that, in 1888, A. Baur discovered that certain nitrobenzene derivatives possessed musk-like odours. Advances in techniques of separation, purification and structural determination enabled more and more of the secrets behind the odour of natural perfume ingredients to be unlocked, and these discoveries paved the way for the synthesis of analogues. Thus, the provision, by serendipity and imitation, of a large palette of perfume chemicals (as opposed to multicomponent natural oils and extracts) with defined odour properties made possible the search for structure–activity relationships (SARs) and the rational design of odorants.

In this chapter, I review firstly some of the theories that have been developed about olfaction, secondly the current state of knowledge regarding the physiology and biochemistry of olfaction, and then I indulge myself in some philosophical considerations regarding mental discipline and scientific method and the apparently too frequent lack of both in approaching an understanding of how our sense of smell works.

THEORIES

Many theories have been proposed in an attempt to explain how humans perceive odours. Those that are still taken seriously can be grouped into two classes, which I label recognition and vibration. In both of these, it is proposed that an odorant molecule comes into contact with a receptor and that the nature of the contact is such that a nerve impulse is generated by the cell containing the receptor. In the recognition theories, the contact is postulated to be that of molecular recognition of a substrate (the odorant) by a protein (the receptor), whilst in the vibration theories it is postulated that the receptor in some way senses a specifiation or group of vibrations in the substrate.

One theory of lesser importance is the radiation theory, first postulated by Aristotle in the fourth century BC. This theory proposes that odorous substances emit radiation, which is detected by the olfactory receptor. In 1947, W. R. Miles and L. H. Beck claimed that bees could detect the odour of honey through a container that was transparent to the far infrared spectrum, but their results have not been repeated. The radiation theory is not given serious consideration nowadays. Other theories that have also been rejected include M. M. Mozell's chromatographic theory (1970), the thermodynamic activity theory, first proposed by P. Gavaudan (1948) and the membrane penetration theory of J. T. Davies (1971). Interestingly, these three all relate to the solubility and/or volatility of the odorant molecule. These properties determine the transport of the molecule from source to receptor and so are important in the overall process though not necessarily in the actual event that generates the signal. Further details of these and other theories are given in the review of Paul Laffort (1994).

The recognition theory was first postulated by Epicurus. In the fourth century BC, he proposed that odorous materials gave off minute particles, which he called atoms, which were detected by the nose. He postulated that smooth, rounded 'atoms' gave rise to sweet odours and sharp, pointed ones to irritant and acidic odours. The best-known modern recognition theory is that of John Amoore (1970) of the US Department of Agriculture in California. He postulated that there is a limited number of receptor types, each of which recognizes a particular molecular shape and, when triggered, generates the signal for a primary odour. The primary odours serve a similar purpose in olfaction to that served by the primary colours of vision. Thus, the enormous variety of odours that we can recognize and describe occurs because of the blending of these primary signals, just as the many hues of colour are composed of appropriate mixtures of red, blue and green. [For a concise account of the current state of knowledge of colour vision, see the article by Hideki Kandori (1995), of Kyoto University.]

Initially, Amoore sought the primary odours by looking for the words that were most commonly used to describe odours. This led him to postulate first seven primary odours, *viz*. ethereal, camphoraceous, musky, floral, minty, pungent and putrid. He studied the chemical and steric properties of typical odorants of each class and proposed shaped receptors for the first five and the generation of charged species for the last two, positive for pungent and negative for putrid. Later on, he postulated that, if these primary odours existed, then specific anosmias should correspond to them. Specific anosmia is the inability of one group of subjects to detect a particular odour. By examining the reactions of a large number of subjects, Amoore attempted to identify anosmias and hence primary odours. This led to an increase in the proposed number of primary odours since, for example, of the four commonest anosmias (musk, sandalwood, ambergris and urine), only one was included in his original set (Amoore, 1970).

There are two other recognition theories which deserve mention. The first is that of M. G. J. Beets (1968), who proposed that the functional group in an odorant serves to align it with the receptor and that the profile of steric bulk thus presented to the receptor is the key determinant of odour. The electron topological approach has been championed by a group of workers at the Academy of Science in Kishinev in Moldavia, principally P. F. Vlad, I. B. Bersuker, M. Yu. Gorbachev and A. S. Dimoglo. In this theory, the recognition involves the electrons of the frontier orbitals of the odorant and the proponents postulated a series of odour triangles. For each of smoke, meat, ambergris and musk, they defined the dimensions of a triangle of atoms and the electronic properties of the orbitals of those atoms, which are postulated to be required for that odour type. In their most recent work on musk, ambergris and sandalwood, they described two molecular fragments of specific composition which must be present in a specific relationship to each other in a molecule for it to possess the given odour.

The vibration theory was first proposed by G. M. Dyson in 1937. He suggested that the receptor in the nose could detect vibrations of the odorant molecules and that patterns of firing of vibrationally tuned detectors could be interpreted as odours by the brain. His theory was taken up by Robert Wright (principally funded by the British Columbia Research Council), who spent a great deal of time in the 1960s and 1970s searching for correlations between infrared spectra and odour. He claimed to have found such correlations, but one of the weaknesses of his theory was that there was no suggestion of how a receptor might sense vibration (Wright, 1982). Such a mechanism has now been provided by Turin (1996). He postulates the presence of an electric potential gap in a protein, with nicotinamide adenine dinucleotide (NAD) and zinc ions providing the 'electrodes'. Electrons cannot cross the gap unless an odorant molecule is placed between the 'electrodes'. To cross the gap the electron must lose energy and this it does by tunnelling through the orbitals of the odorant molecule and exciting vibrational modes in it as a result. Thus, Turin has moved the search for correlations from infrared to inelastic electron tunnelling spectra.

All of these theories of chemoreception have sprung up largely in the absence of hard biological information and are based on ideas of how it might work. They are based on observations of the total process, which is a dangerous thing to do since there are many stages between the evolution of an odour from its source and the consciousness of odour in the brain. The progress at each stage involves an interplay of physical and/or chemical and/or physiological and/or psychological parameters. Models based on the above theories usually adopt a very mechanistic approach and largely ascribe discrimination to one stage in the process, *viz*. the odorant–receptor interaction.

BIOLOGICAL FACTS

In humans, odorants are detected by the olfactory epithelium. This is a patch of yellow tissue located at the top of the nasal cavity, approximately on a level with the eyes. Two other organs that are thought to be involved in olfaction are the trigeminal nerves on the side of the cavity and the vomeronasal organ on its base (Figure 13.1). The trigeminal nerve seems to be involved in the detection of irritants. The role of the vomeronasal organ is the subject of considerable controversy, and it could be vestigial in humans. Nerve signals from the epithelium are relayed through the cribriform plate into the olfactory bulb and from there to the limbic system and on to the higher brain.

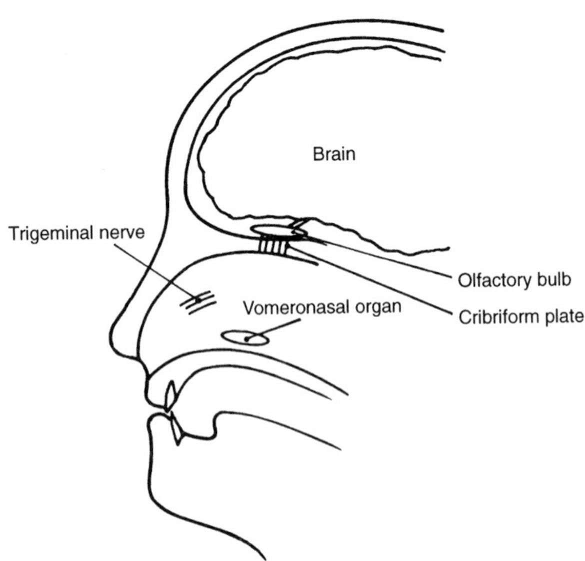

Figure 13.1 *Key organs thought to be involved in the perception of odour*

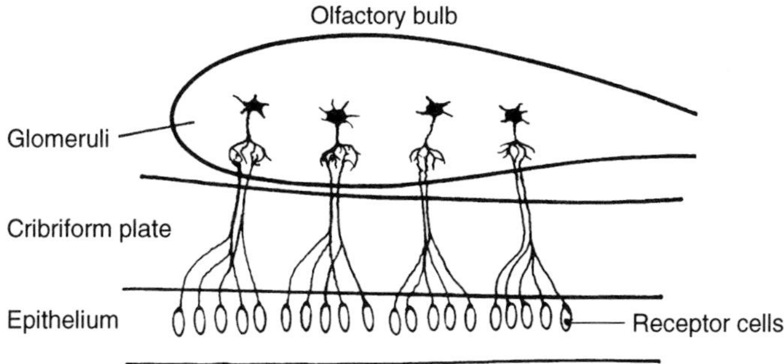

Figure 13.2 *Convergence of receptor cell signals on to glomeruli in the olfactory bulb*

The neurons from the epithelium converge on areas of the olfactory bulb known as glomeruli. Evidence suggests that all of the signals from one type of receptor converge on a common glomerulus (Figure 13.2).

Figure 13.3 shows the receptor cell and some of the molecules associated with it. Odorants from the air in the nose enter the mucus layer around the receptor cell. It is uncertain how they cross the mucus to reach the receptor cell. The process could simply be diffusion, which is possible as odorant molecules usually have some degree of water solubility, although hydrocarbons such as iso-octane do have odours.

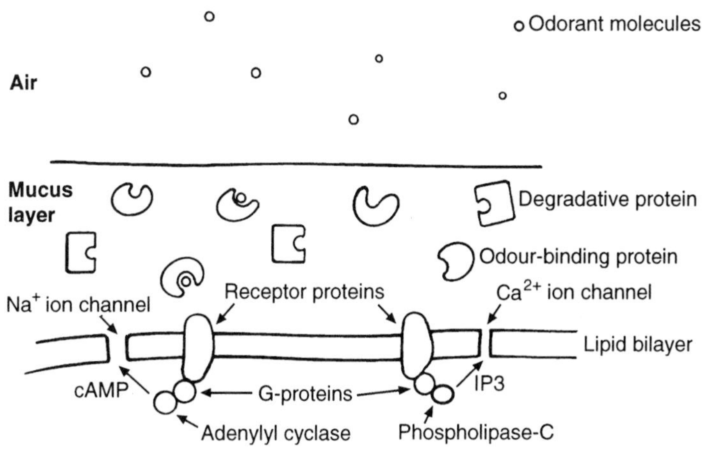

Figure 13.3 *Molecules known to be involved in generation of an electronic signal by odorants*

Another possible mechanism is the transport of odorant molecules across the mucus by odour-binding proteins (OBPs). These proteins are known to be present in the mucus and are known to bind odorants. It is possible that the OBPs could deliver the odorants to the receptor proteins or that the receptor proteins actually bind to an OBP–odorant complex. The former is the more generally accepted hypothesis. In either case, it is possible that the OBPs are involved in recognition since there are distinct classes of OBP and their affinities for odorants and for different receptor proteins are expected to be different.

Degradative proteins are also present in the mucus, the role of which is presumably to remove odorants continually and therefore ensure that signal generation ceases when the source of the odour is removed. The receptor proteins sit in the lipid membrane of the cell. The backbone of the protein passes through the membrane seven times and so one terminus is outside and one inside the cell. The portions that pass through the membrane exist as α-helices. These α-helices come together to form a cylinder and it is thought that the binding sites lie inside this cylinder. This is consistent with hormone receptors and with optical receptors in which the retinal-derived pigment is held in such a way. Two groups of workers have studied potential binding sites in the rat olfactory receptor protein *OR*5. One identified a possible binding site for Lilial® [2-methyl-3-(4-*t*-butylphenyl)propanal], and the other a potential site for hydroxycitronellal. The two binding sites are in similar areas of the protein. In both cases, the work involved molecular modelling rather than isolation and characterization techniques, and the evidence is therefore circumstantial. However, it is interesting to note that Lilial® and hydroxycitronellal have similar muguet (lily of the valley) odours.

Interaction of the odorant with the receptor protein causes a change in the structure of the latter (by hydrogen bonding, electron tunnelling or some other means), which causes the G-protein associated with the receptor to activate either phospholipase-C or adenylyl cyclase. In the former case, the second messenger is inositol triphosphate (IP3) and in the latter it is cyclic adenosine monophosphate (c-AMP); both are released into the cell. These second messengers cause calcium and sodium ion channels to open in the cell membrane and thus generate nerve signals. The second messengers also set in train a process that deactivates the receptor by phosphorylation, and a dephosphorylation is necessary to make it active again.

The gene family that encodes the receptor proteins is the largest in the genome. Potentially, there are thousands of different receptor proteins. However, not all are expressed in any one species and those that are expressed in one animal are not expressed uniformly across the epithelium. Thus, information about the nature of the odorant could also be gained from a knowledge of the intensity of signals arising from different areas of the epithelium. The receptors expressed in fish are from a different part of the 'family tree' of receptors from those expressed by mammals. Intriguingly, the nasal cavity of the frog is divided into four sections rather than the two (left and right) of mammals. One on each side of the frog's head is open to air when the frog breathes and the other two are exposed to water when the frog is submerged. A valve switches between the two states. The receptor proteins expressed in the 'wet' nose are similar to those of fish and those in the 'dry' nose to those of mammals. Tadpoles start life with only the 'wet' nose and the 'dry' nose develops during metamorphosis into frogs.

Further information on the biochemistry of olfaction can be found in the excellent review by Heinz Breer *et al.* (1994) of Hohenheim University, Stuttgart.

MISCONCEPTIONS

Odour has three properties, *viz*. character, intensity and persistence. All three are subjective and can only be measured in sensory terms. Most correlation work has been done on character since it is, superficially, the easiest to measure. However, the description of an odour is associative, since we have no hard reference points. Difficulties in finding correlations can arise from the use of odour classification systems. For instance, in our fragrance work, we found that the use of the term fruity to describe an odour family led to confusion, since the criteria for a molecule to possess an apple odour are not the same as those for pear. We classify the two together because of the similarity of the botanical sources, but this is not necessarily related to the odour properties. To study structure–odour correlations, we must therefore ensure that we are using meaningful parameters.

The interaction between the study of the mechanism of odour perception and the elaboration of empirical SARs has led to the problem of confusion of the two. It is only too easy to fall into this trap and the lessons which can be learned from the study of odour are, I believe, relevant to most, if not all, branches of science.

It is always tempting for the chemist to interpret observations on SARs in mechanistic terms and to postulate means by which the biological activity could be brought about through some feature of molecular structure or chemical activity which seems significant in the

SAR. It is also tempting to do the reverse; that is, to postulate a biological mechanism and then work backwards to structural features that might satisfy it. Both approaches have been tried in the field of olfaction and, to date, without unequivocal success. The development of a biological model has the attraction of providing a simple approach to the rational design of active molecules. However, it does tend to lure one into forgetting the basic tenets of the scientific method and going somewhat astray in consequence.

Science advances by the formulation of theories. The scientific method comprises observation of the facts, hypothesis of a theory which accommodates all of the facts and then testing the theory by using it to predict results which are unknown at the time. No theory can ever be proved, it can only be disproved and so is valid only as long as there are no known exceptions. The platitude 'The exception that proves the rule' is misunderstood by most of those who use it. They forget that here the word 'proves' is being used in its original sense of 'tests' rather than in the modern sense of 'establishes beyond doubt'. Any exception means that the rule is either inadequate or invalid and must be revised or rejected. As far as theories concerning the initial receptor event in olfaction are concerned, none accommodate all of the known facts.

Another common failing in the logic behind odour theories is the inability to distinguish between cause and effect. All too often, someone finds a correlation between two parameters and assumes a causal relationship, without asking whether this correlation could simply be between two effects of a common cause. An example of this in the field of olfaction is the assumption that a correlation between the infrared spectra of a set of odorants and their odours demonstrates that the odour was caused by those specific vibrations. The odours and the spectra are both effects of a common cause, that is, the molecular structure.

Structure–odour correlations rely on the measurement of odour. It must be remembered that odour is not a physical property. Odour exists in the mind. We must not assume that it exists in the receptor also. The fact that cross-adaptation can occur between chemically different materials with similar odours, and also between chemically similar materials with different odours, suggests that it does not. There is no evidence that the receptors are tuned to specific odours or even odour types. It is quite likely that the recognition process is based on physical and/or chemical properties of odorants and that odour only exists when the brain puts an interpretation onto a pattern of signals from the olfactory nerves.

For example, one of the famous problems of structure–odour correlation is that of the odour of bitter almonds. The two compounds that best elicit this odour are benzaldehyde and hydrogen cyanide. Examination of analogues and substitutes for benzaldehyde show that the more closely a molecule resembles benzaldehyde in stereoelectronic terms, the more likely it is to smell of almond. But, of course, hydrogen cyanide is quite different from benzaldehyde. If we assume that there is an almond receptor, we are faced with a very difficult problem in accommodating the facts. However, if we think of the almond odour as something which exists only in the brain as a result of its interpretation of a pattern of nerve signals, then there is no difficulty. The interpretation is learnt, and the brain has learnt to interpret two different sets of signals (those arising from benzaldehyde and those from hydrogen cyanide) in a similar way since, in the smell of natural almonds, the two molecules occur together as the result of the breakdown of the flavour precursor, amygdalin (the glycoside of benzaldehyde cyanohydrin).

Equally, what we term odour is, in fact, the end result of a number of discrete processes. In consequence, we cannot assume that any correlation between agent and effect arises from only one of these. All too often it is assumed that only the interaction between the odorant and the receptor protein is of significance and that structure–odour correlations give mechanistic information about this specific event.

The general lesson to be learnt from these two mistakes is that we can only interpret SARs, or indeed any other correlations, in mechanistic terms if we are certain that we are measuring the direct input and output from a single process. Of course, the correlation might be a valid and useful tool even if it does not give mechanistic information.

CONCLUSIONS

We are still a long way from knowing all the details of the process of olfaction and odour discrimination. I believe that it is more complex than most workers have previously imagined. I think that we will discover that there are many different layers of discrimination. For example, binding proteins, geography of the epithelium and pathways in the nerves might all be involved in the overall recognition process. If different binding proteins show different selectivities towards both odorants and receptor proteins, then, even if the discrimination at each stage is only based on a few broad bands, there would still exist a large number of possible combinations and hence a wide range in signal patterns. If I am correct, then understanding the mechanism of olfaction will be unlikely to help in the prediction of odours for a very long time and the fragrance industry will therefore have to continue to rely on empirical SARs. However, we cannot know that this is the case until we have done the necessary research. This work covers a wide range of disciplines and it is most unlikely that the solution will be found by the old approach of one person working in isolation.

On a more general note, the mistakes made by workers in olfaction, myself included, illustrate the need to keep to the scientific method, to maintain clarity of logical thinking rigorously, to question assumptions and to escape from paradigms, when faced with large, complex problems.

REFERENCES

J. E. Amoore, *Molecular Basis of Odour*, Charles C. Thomas, 1970.

H. Breer, K. Raming and J. Krieger, Signal Recognition and Transduction in Olfactory Neurons, *Biochim. Biophys. Acta*, 1994, **1224**, 277–287.

H. Kandori, *Chemistry and Industry*, 1995, **18**, 735.

P. Laffort, Relationships between Molecular Structure and Olfactory Activity, in *Odours and Deodorization in the Environment,* G. Martin and P. Laffort (eds), VCH, 1994.

L. Turin, A Spectroscopic Mechanism for Primary Olfactory Reception, *Chem. Senses*, 1996, **21**(6) 773–791.

R. H. Wright, *The Sense of Smell*, CRC Press, 1992.

M. G. J. Beets, Odour and Molecular Structure, *Olfactologia*, 1968, **1**, 77.

G. M. Dyson, Raman Effect and the Concept of Odour, *Perf. Essent. Oil Rec.*, 1937, **28**, 13.

I. B. Bersuker, A. S. Dimoglo, M. Yu. Gorbachov, M. N. Koltsa and P. F. Vlad, *New J. Chem.*, 1985, **9**, 211.

I. B. Bersuker, A. S. Dimoglo, M. Yu. Gorbachov and P. F. Vlad, *New J. Chem.*, 1991, **15**, 307.

FURTHER READING

R. Axel, The Molecular Logic of Smell, *Sci. Am.*, 1995, October, 130–137.

K. J. Rossiter, Structure–Odor Relationships, *Chem. Rev.*, 1996, **96**(8), 3201–3240.

Chapter 14

Electronic Odour Sensing

JENNY OLIVER

INTRODUCTION

It is well known that many people have some 'gaps' in their olfactory capability, known as 'specific anosmias'. Furthermore, the human nose tends to 'tire', or become desensitized, after a short period. Aromas detected initially begin to fade in one's consciousness after varying lengths of time, dependent upon the odours involved. It is not difficult, therefore, to imagine an instrument which is the 'perfect nose': one that detects every odour, does not become desensitized and always give the same description to a specific odour. Unfortunately, it is not that simple!

Human noses do need to 'tire': imagine walking off the busy street into a chemist's shop. One would immediately detect the odour of the perfumes, fragranced toiletries and other goods, but the outside smells would be very much in the background. However, an instrument just inside the door would still be picking up the exhaust and diesel fumes of the passing cars and buses. In many applications of an ideal artificial nose, this needs to be catered for. Currently, the only answer is to ensure no extraneous odours or non-odorants enter the machine.

Several instruments have emerged in the marketplace and have been popularly described as 'electronic noses'. In reality, these are volatile chemical sensor arrays. To give them more likeness to the human nose (or, more accurately, olfactory system), they are coupled to artificial intelligence systems that require development. The instruments currently available detect most vapours, odorous and non-odorous, including water vapour (to which they are all highly sensitive). Sensitivity to other volatiles, however, is highly variable, dependent upon the type of sensor used. Some airborne materials that are very easily detected by a human nose cannot yet be detected and/or identified satisfactorily by the instruments available at the time of writing; for example, very low levels of some sulfurous compounds and steroids.

ELECTRONIC AROMA-SENSING SYSTEMS

Instruments from the major manufacturers utilize thin-film technology. Semiconducting metal oxides (doped tin) and semiconducting polymers comprise the majority (Figures 14.1–14.3). These work using principles of a change in resistance within the layer when volatiles are adsorbed on to the surface (Gardner and Bartlett, 1992). The other principal sensor type is the mass-sensitive resonating quartz crystal. This usually uses a surface acoustic wave (SAW, a high-frequency surface standing wave), the frequency of which is depressed as the molecules of the volatiles are adsorbed onto the film's surface (Wünsche *et al.*, 1995). Less well-developed technology includes the use of thermo-resistive microcantilevers, which operate rather like microscopic bimetallic strips (Gimzewski *et al.*, 1994). Many other types of chemical sensor

Figure 14.1 *The AromaScanner®*

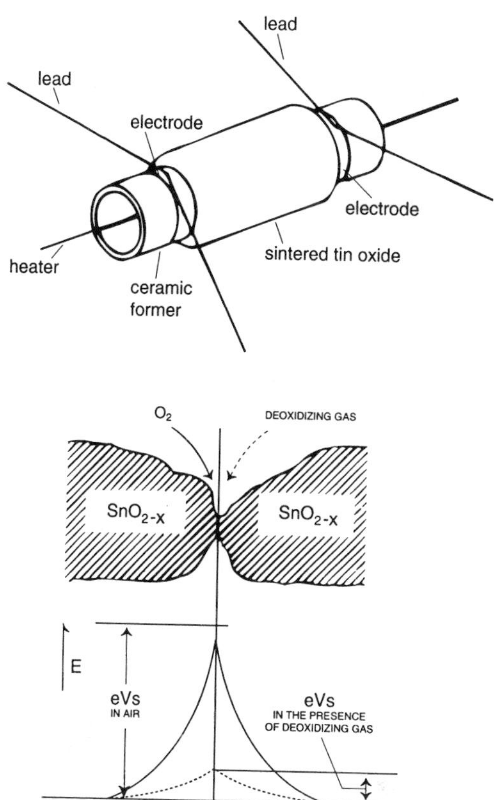

Figure 14.2 *AlphaMOS metal oxide sensors*

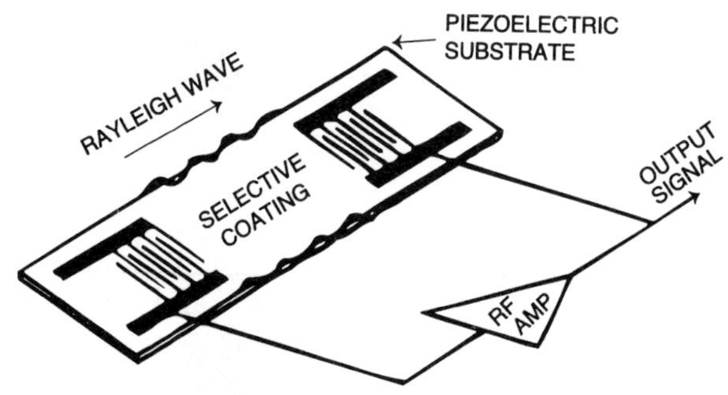

Figure 14.3 *Surface acoustic wave sensor*

may be incorporated into an electronic nose in the future. A review document (Janata *et al.*, 1994) published in *Analytical Chemistry* provides an excellent introduction to the types available. One of the most recent sensor types (at the time of writing) to be incorporated into an experimental artificial nose is based on fluorescence responses and optical fibre links (Dickinson *et al.*, 1996). Chemo-optical sensors generally appear to be gaining interest, and comprise the largest number of references in the review (Janata *et al.*, 1994).

One of the more widely accepted theories on olfaction suggests that odour character is related to the shape and functionality of the molecule (Amoore, 1962; Beets, 1961). Although it is not known precisely how the thin-film sensors operate, some principle relating to the surface charge and topography of the odorant molecules is operational when they are adsorbed on to the film surface. The sensors utilized in artificial noses, particularly the semiconducting polymer types, are generally of low specificity; that is, they react to a wide range of molecules. However, each sensor in an array reacts at a different level. The result, after selecting a representative averaged 'time slice', is an 'odour fingerprint'.

The instruments have been compared to a musician with a manuscript. If we see the musical score of, say, Sibelius's *Finlandia*, it might be difficult to imagine the associated sound unless we are musicians. Electronic odour sensing seeks to transcribe, as it were, a sensory stimulus for one sense (olfactory) to another (visual). In addition, this 'transcription' is required to be measurable and sufficiently reproducible to allow comparison. The analogy with music continues. Just as with a musical score, the transcription cannot reproduce all of the attributes; for example, the precise speed of performance, the balance of the instruments and so on; only artificial reproduction in the form of a recording can do that. In the same way, electronic odour sensing can only depict some of the attributes of a fragrance.

The semiconducting systems tend to be less sensitive to molecules of higher molecular weight, say molecules of 15 carbon atoms and above. The mass-sensitive SAW system is perhaps more appropriate where measurement of these molecules is required, such as airborne steroids in body-odour research. Indeed, there is a hybrid machine on the market which utilizes metal oxides, semiconducting polymers *and* SAW technology. However, all commonly available sensors used in commercial electronic noses are still sensitive to water vapour to some extent, although the problem has been addressed by various means. One solution is the use of a programmable integral differential (PID) controller for the reference air humidity. This is particularly useful for the sensor arrays that do not function so well in nitrogen or dry air. Other solutions to this problem involve the use of dry purge gases in a sample chamber, but this is further removed from replication of an 'everyday' human nose.

The artificial intelligence systems to which sensor arrays are coupled supply the closest likeness to the human olfactory system. Some of the recent theories on olfaction require that the human nose has only relatively few types of receptor, each with low specificity. The activation of differing patterns of these receptors supplies the brain with sufficient information for an odour to be described, if not recognized. As a consequence of this belief, the volatile chemical-sensing systems commercially available only contain from 6 to 32 sensors, each having relatively low specificity. Statistical methods such as principal component analysis, canonical discriminant analysis and Euclidian distances are used for mapping or linked to artificial neural nets as an aid to classification of the odour 'fingerprints'.

POSSIBLE USES OF AN ELECTRONIC ODOUR-SENSING SYSTEM

The forte of these instruments appears to be their classification capability. An appropriately trained system can classify odours into 'like', 'unlike' and varying degrees of similarity in between. The underlying nature of this similarity is not indicated by the analysers and has to be inferred with a structure–property–activity analysis by the experimenter (see Chapter 11). The classification can allow one to infer 'pass' or 'fail', together with reasons for failure if appropriately trained. This is probably the area for most progress, once the humidity problems have been addressed satisfactorily. A library of data for, say, natural oils could yield a method analogous to olfactory quality control, although it must be recognized that the library for one instrument may not be usable by another.

The sensors are rather like a set of different chromatographic stationary phases. They each detect almost every volatile to a differing degree, including commonly utilized solvents. Many of the solvents used in perfumery have little or no odour. A difference in dilution (or diluent) undetectable by a human olfactory system could cause a large difference in the odour 'fingerprint' when using an electronic device. This is partly because the sensor reaction is more nearly linear with respect to concentration in the air. The human nose appears to have logarithmic characteristics once the odorant concentration has achieved a certain threshold.

Apart from the use as a Quality Control device, other areas in which an artificial odour-sensing system could be utilized include all those in which classification of odour is required; for example, human body odour, malodours and malodour counteractancy. Another area in which the new instrumentation could be utilized to advantage includes perfume substantivity, or diffusion from a substrate. For example, it could be used to measure levels of perfume in the air from a hard surface cleaner when used on a ceramic tile, or odour from human skin after spraying with a cologne, and so on.

CONCLUSION

The novel ranges of electronic odour-sensing devices offer an exciting prospect to the world of perfumery. It must be remembered, however, that the technology is still in its infancy. Nevertheless, it is not so many years ago that some gas chromatographs started life as a heated chamber large enough to walk in, and contained a column packed with brick dust! At the time of writing, novel aroma-sensing systems are already further ahead than that. If we consider where gas chromatography is today, the next few years should prove an exciting time in the field of electronic odour-sensing devices.

REFERENCES

J. E. Amoore, The Stereochemical Theory of Olfaction, *Proc. Sci. Sec. Toilet Goods Assn.*, 1962, **37**, Parts 1 & 2 (1–23, special supplement).

M. G. J. Beets, Odor and Molecular Constitution, *Am. Perfumer*, 1961, **76**(6), 54.

T. A. Dickinson, J. White, J. S. Kauer, and D. R. Walt, A Chemical-Detecting System Based on a Cross-Reactive Optical Sensor Array, *Nature*, 1996, **382**, 697–700.

J. W. Gardner and P. N. Bartlett (eds), *Sensors and Sensory Systems for an Electronic Nose*, Kluwer Academic Publishers, 1992.

J. K. Gimzewski, Ch. Gerber, E. Meyer and R. R. Schlittler, Observation of a Chemical Reaction using a Micromechanical Sensor, *Chem. Phys. Lett.*, 1994, **217**, 5–6.

J. Janata, M. Josowicz, D. M. DeVaney, Chemical Sensors [Review document with over 700 references], *Anal. Chem.*, 1994, **66**(12), 207R–228R.

L. F. Wünsche, C. Vuilleumier, U. Keller, M. P. Byfield, I. P. May and M. J. Kearney, Scent Characterisation: from Human Perception to Electronic Noses, in *Flavours, Fragrances and Essential Oils*, Proc. 13th International Congress of Flavours, Fragrances and Essential Oils, K. H. C. Baser (ed.), 1995, vol. 3, p. 295.

Chapter 15

The Search for New Fragrance Ingredients

KAREN ROSSITER

INTRODUCTION

The perfumers of today have at their disposal over 3000 fragrance ingredients to choose from when creating a perfume. These include natural oils, nature-identical materials and synthetic compounds (see later for definitions). With such a large number to choose from, it might at first be thought that the perfumer does not need any more. In this chapter, the stability of a range of lily of the valley (muguet) odorants is used to illustrate why the introduction of new fragrance ingredients is important. Other reasons for extending the perfumer's palette are also mentioned. Continuing with the muguet theme, examples are given to explain how a chemist might go about the search for novel materials. The more traditional approaches, such as the analysis of natural products, serendipity and lead optimization, are illustrated only with one or two examples, since these techniques have been discussed in more detail in Chapters 4 and 12. Instead, the use of structure–activity relationships (SARs) is examined in more detail for the following reasons. First and foremost, this technique can be used to predict the activity of compounds that have not yet been made and as such can lead to the discovery of new active compounds by design rather than by chance. Second, advances in theoretical and computational chemistry and the introduction of more powerful computers have resulted in the rapid development of more sophisticated SAR techniques. Three SAR approaches are discussed: the Hansch analysis, the osmophore approach and pattern recognition.

There are two unique problems in the application of SAR to odour. The first of these is the difficulty associated with odour measurement (see Chapter 8 for details on the measurement of odour perception). This difficulty arises from the subjectivity of odour, the dependence of perceived odour strength and character on concentration and the importance of organoleptic purity as opposed to chemical purity. All of these factors can contribute towards discrepancies in odour data from different sources. A useful technique for checking the olfactory purity of a sample is gas chromatography–olfactometry (GC–O), more commonly referred to as GC-sniffing (see Chapter 12). Provided that the GC conditions adequately separate the components of a mixture, each component can be smelled in an olfactory pure state at the exit port of a GC column. It is not uncommon for an odour to be perceived at a position in the gas chromatogram where there is no peak. This illustrates the vastly superior sensitivity of the human nose over even today's most sensitive GC instruments, and shows that even the smallest trace of a strongly odoriferous material can alter the odour profile of a sample completely. The problems associated with subjectivity are minimized by the use of an expert panel and a standard glossary of odour descriptors. Odour data obtained in this way are the most reproducible that one can obtain, provided, of course, that the samples are organoleptically pure. It should be remembered that without good, precise, reproducible data all subsequent SAR work is a waste of time.

The second unique feature is that our understanding of the mechanism of olfaction is presently rather limited. Prior to the early 1980s, the biochemical study of olfaction had been virtually neglected. However, over the past 15 years a great deal of information has been gathered on olfactory receptors and odour transduction, which has provided us with some initial insights into the biochemistry of odour perception. A detailed account of the theories and biochemical study of olfaction is outside the scope of this chapter. For further reading on these topics, see the articles of Breer *et al.* (1994), Laffort (1994), Pelosi (1996), Ronnett (1995) and Sell (1997). Current research activities in this area are likely to lead to an increase in our understanding of the olfactory system, which in turn should make the search for correlations between structure and odour easier. Already we are starting to see publications describing the molecular modelling of interactions between fragrance ingredients and putative olfactory receptors. Although modelling of this class of protein is at present highly speculative, it could well be the ingredient design technique of the future, and therefore a brief section on ligand–receptor modelling is included at the end of the chapter.

THE NEED

The number of ingredients available to a perfumer creating a fragrance for relatively cheap and hostile products, such as strongly alkaline bleach or acidic antiperspirant, is significantly lower than the number available for use in alcoholic fine fragrances. The perfumer's freedom of choice is considerably restricted by factors such as stability and cost. For example, Figure 15.1 shows the relative organoleptic stability of a range of muguet ingredients (1–6) in an aerosol antiperspirant. The organoleptic stability of an ingredient in a product base is determined by comparing the odour of the test sample, which has been stored at a specific temperature (usually 25 or 37 °C) for a given number of weeks (4, 8 or 12), with that of a control and ranking it as very good, good, moderate or poor. The control is either a freshly prepared mixture of the fragrance ingredient in the product base or a sample which is as old as the test sample, but which has been stored at a lower temperature to minimize deterioration of the fragrance ingredient. When the odour of the test sample and the control are virtually identical, the fragrance ingredient is said to be organoleptically stable and it is ranked as very good. If either the odour of the fragrance ingredient has virtually disappeared to leave only the odour of the product base or the sample has developed undesirable olfactory notes, the fragrance ingredient is classified as having poor organoleptic stability. In Figure 15.1 the organoleptic stability of the six muguet ingredients is represented by a bar diagram: the taller the bar, the more stable the ingredient. It is clear from this figure that in creating a muguet-type fragrance for use in an

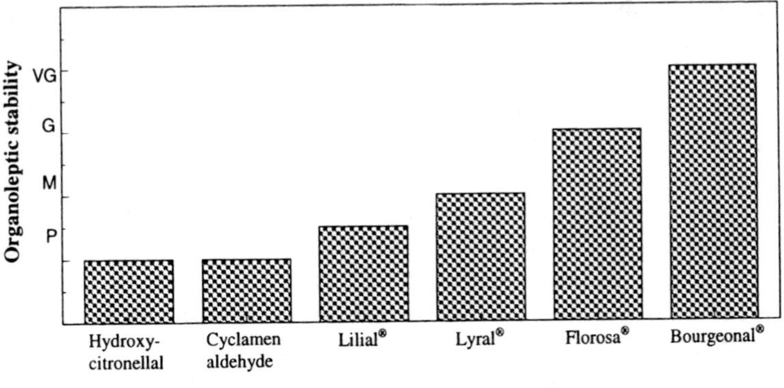

Figure 15.1 *Organoleptic stability of various muguet ingredients in an aerosol antiperspirant after storage at 37 °C for 12 weeks (VG = very good; G = good; M = moderate; P = poor)*

aerosol antiperspirant, the perfumer is forced to use either Florosa® (5) or Bourgeonal® (6).

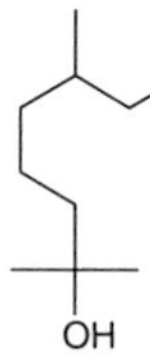

Hydroxycitronellal
(**1**)

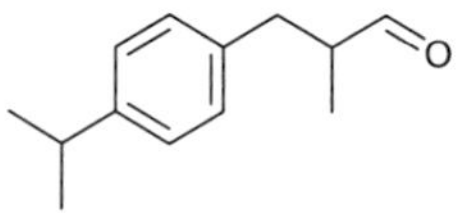

Cyclamen aldehyde
(**2**)

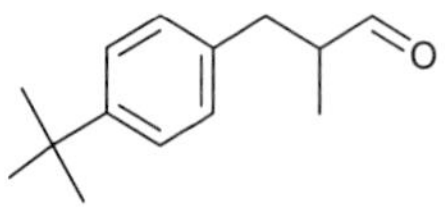

Lilial® (Givaudan)
(**3**)

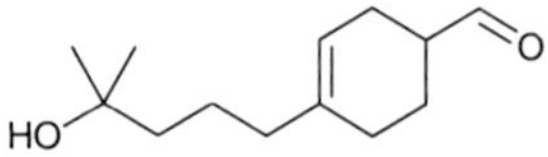

Lyral® (IFF)
(**4**)

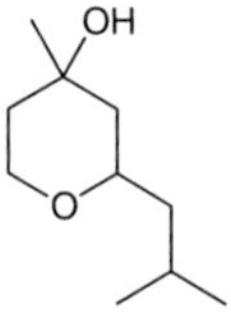

Florosa®
(Quest International)
(**5**)

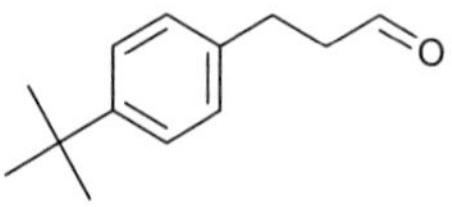

Bourgeonal®
(Quest International)
(**6**)

The organoleptic stability of an ingredient depends on a combination of two factors: its sensory characteristics and its chemical stability. The latter, of course, is determined by the nature of the product base. Antiperspirant formulations are acidic because of partial hydrolysis of the active antiperspirant agents, such as aluminium chlorhydrate (equation 1). It is an inorganic salt that consists essentially of complex aluminium chloride described empirically as $[Al_2(OH)_5]_n.nCl$. The complex is polymeric and loosely hydrated.

$$[Al_2(OH)_5]^+ + xH_2O \rightleftharpoons [Al_2(OH)_{5+x}] + xH^+ \tag{1}$$

x is a variable depending on concentration

Florosa® (5) performs well in antiperspirants because it is chemically more stable than the aldehydes. Ether groups are relatively inert and, although possible reactions of the alcohol group depend upon the specific formulation of the antiperspirant, there is relatively little it could do other than acid-catalysed dehydration, which usually requires fairly high temperatures. Consequently, the level of Florosa® in, for example, an ethanol-based roll-on antiperspirant after 4 weeks storage at 37 °C remains unchanged. In contrast, the percentage of Lilial® remaining is only 15%. This huge difference arises from the reactivity of the aldehyde functional group, which reacts with ethanol to form the corresponding diethyl acetal (8) and undergoes autoxidation to yield the corresponding carboxylic acid (13). The acetal formation is acid-catalysed and proceeds via the hemiacetal (7) (Scheme 15.1).

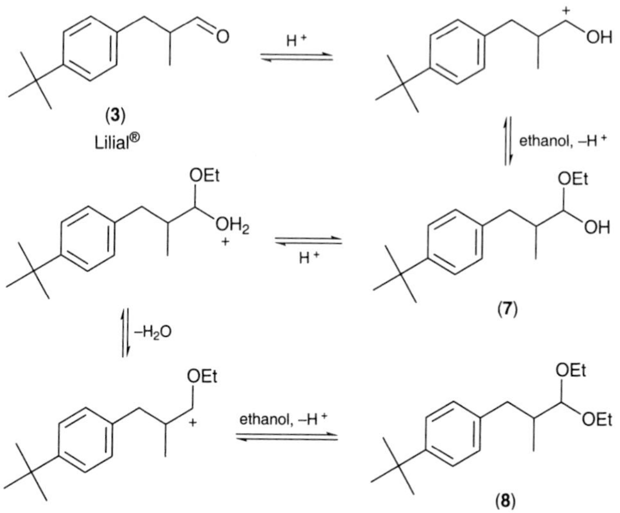

(3)
Lilial®
H^+
ethanol, $-H^+$
OEt
OH
(7)
OEt
OH_2^+
H^+
$-H_2O$
OEt
ethanol, $-H^+$
OEt
OEt
(8)

Scheme 15.1

Autoxidation is defined as the reaction of organic compounds with oxygen under mild conditions (Scheme 15.2). This chemical transformation is so facile that if, for example, a thin film of Lilial® is placed in a dish overnight the liquid is converted into a white crystalline solid, the corresponding carboxylic acid, by the morning. Oxygen itself is too unreactive to be the species which actually abstracts the aldehydic hydrogen atom. However, if a trace of free radical (9) is produced by some initiating process, it reacts with oxygen to give the peroxy radical (10). This radical removes a hydrogen atom from a second molecule of Lilial® to form the peracid (11) plus the acyl radical (9). The peracid (11) reacts with a further molecule of Lilial® to form the adduct (12), which then decomposes to give two molecules of acid (13).

In the case of Lilial®, and other aldehydes that have an alkyl substituent in the α-position, there is a second possible autoxidation process. The acyl radical (9) can be decarbonylated to give the secondary carbon radical (14), which can react with oxygen to form the corresponding hydroperoxide (15). Under acidic conditions the hydroperoxide is protonated (16) and a hydrogen atom migrates to form the alkoxycarbonium-ion intermediate (17), which on hydrolysis yields *p*-*t*-butylphenylpropan-2-one (18). Out of the autoxidation

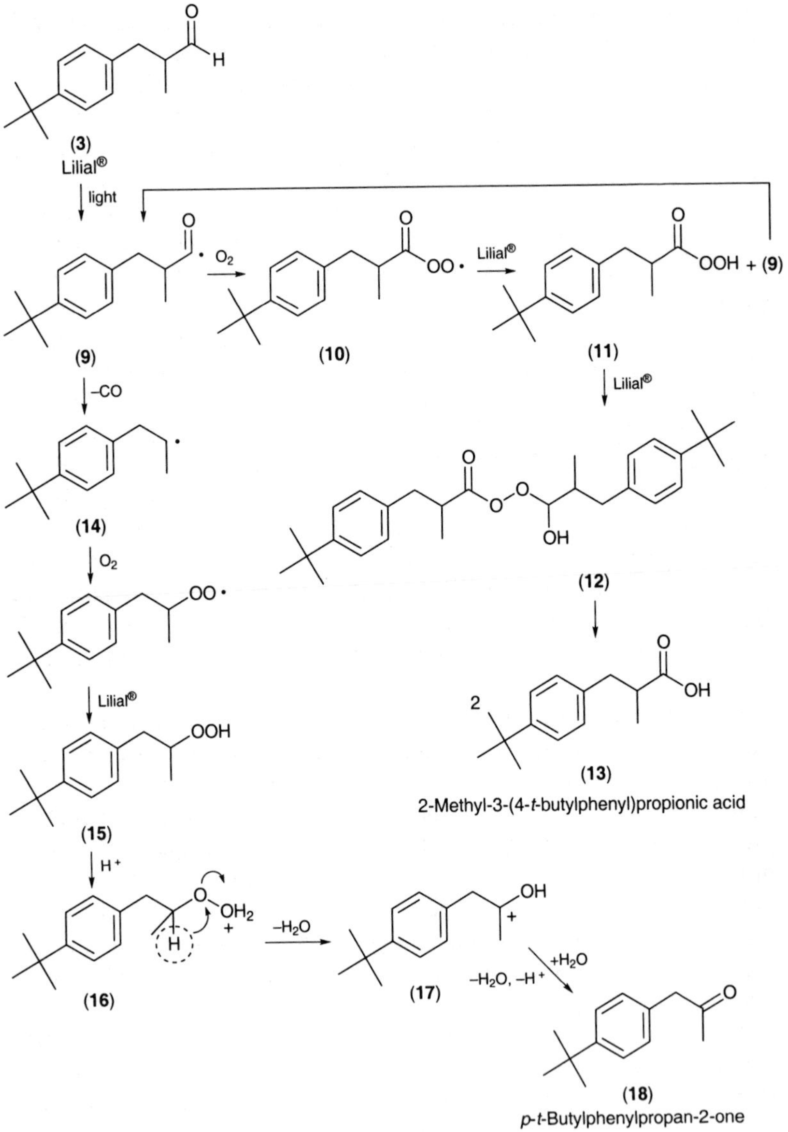

(3)
Lilial®

light

O_2

Lilial®

OOH + (9)

(9)

(10)

(11)

Lilial®

–CO

(14)

(12)

O_2

Lilial®

(13)
2-Methyl-3-(4-*t*-butylphenyl)propionic acid

(15)

H^+

$-H_2O$

(16)

(17)

$+H_2O$

$-H_2O, -H^+$

(18)
p-*t*-Butylphenylpropan-2-one

Scheme 15.2

processes, it is formation of the carboxylic acid that predominates, with typical ratios of compound 13 to compound 18 being about 20:1.

The initial step of both the above reactions, the formation of an acyl radical, is catalysed by light and by metal ions that are capable of a one-electron reduction transition (*e.g.* $Fe^{3+} \rightarrow Fe^{2+}$). Thus, the autoxidation of aldehydes can be greatly slowed by keeping the compounds in the dark and by very careful purification. However, the most efficient method is the addition of antioxidants, such as phenols and aromatic amines, which react preferentially with any radicals that may be present.

Bourgeonal® (6) performs well in antiperspirant for a combination of reasons. Our studies have shown that, despite its structural similarity to Lilial®, it is substantially more stable. After 4 weeks at 37 °C the level of Bourgeonal® remaining in an ethanol-based roll-on is typically 65%, as opposed to only 15% for Lilial®. Although like Lilial®, Bourgeonal® reacts with ethanol to form the corresponding diethyl acetal, it does not appear to undergo autoxidation. The reasons for this remain unclear. The only structural difference between these two compounds is the absence of a methyl group in the alpha position of Bourgeonal®. Although this inhibits autoxidation to the corresponding lower homologue ketone because of the lower stability of primary radicals as opposed to secondary ones (14), it would still be expected to undergo autoxidation to the corresponding carboxylic acid. It is well known that electron-donating groups, such as a benzene nucleus and an ether oxygen atom, facilitate abstraction of the aldehyde hydrogen atom to form the initial acyl radical (9). However, the effect of a methyl group is going to be so small that it does not explain the huge differences observed between autoxidation of Lilial® and Bourgeonal®. The second reason for the good performance of Bourgeonal® is that it is a very potent material, being 2–4 times stronger than Lilial® at the same concentration (Boelens and Wobben, 1980; The Givaudan Corp., 1961).

Although Florosa® is more stable than the aldehydic muguet materials, it cannot be used as a simple replacement in a formulation. Its floral odour is more rosy and it lacks both the impact and the green and watery notes associated with materials such as Lilial® and Bourgeonal®. Ideally, the perfumer needs a new fragrance ingredient with the chemical stability of Florosa® and the odour properties of Lilial® and Bourgeonal®. This is the challenge for the fragrance chemist. However, even if a suitable molecule could be found today, it is unlikely that it will be commercially available to the perfumer for another 3–5 years. It takes this long for a new compound to clear all of the regulatory, safety, process and production hurdles. Thus, the discovery chemist's role is proactive. Problems tackled 'today' will be of enormous value to the perfumery industry of 'tomorrow'. In the meantime, the perfumer working on today's antiperspirant fragrance for the Business Scents Ltd brief is guided by olfactory stability data. The Lilial® in the original formulation will be replaced with a cheap, stable alternative such as Florosa® and the perfumer's skills in creative perfumery used to achieve the desired impact and green, watery character (Chapter 7).

The above example illustrates just one of the reasons why there is an ongoing need to extend the perfumer's palette. Other advantages that new ingredients might offer over existing aroma chemicals include better olfactory properties, a lower production cost, improved biodegradability, uniqueness of odour and additional functionality (*e.g.* a fragrance ingredient that also repels insects). A novel aroma chemical is usually kept 'captive' for a few years, *i.e.* Quest International will only manufacture the compound for internal usage. This gives our perfumers a creative advantage over those of our competitors.

THE SEARCH

Where does a chemist begin the search for a new fragrance ingredient? Well, several options are available, each of which is described below using lily of the valley (muguet) as an example.

Clues from Nature

The unprecedented development of analytical techniques from the 1960s through to the present time has enabled the identification of key components in natural products (see Chapter 12) and their subsequent synthesis in the laboratory. Compounds that are prepared by chemical means but that are identical in structure to those found in nature are known as nature-identical materials. The synthesis of compounds closely related to these nature-identical materials has led to the discovery of ingredients that are similar in odour, but which are much easier and cheaper to make than their natural counterparts. A notable example is that of jasmine: jasmine absolute costs £3000–5000/kilogram, the nature-identical materials (19) and (20) cost £300–500/kilogram and the simpler cyclopentanone derivatives (21–23) £10–50/kilogram.

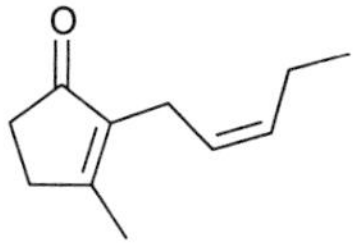

(**19**)
Jasmone

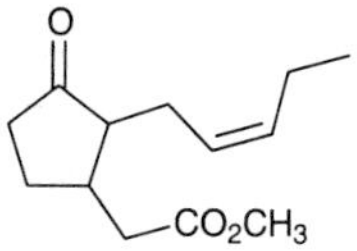

(**20**)
Methyl jasmonate

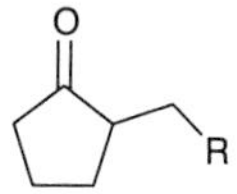

(**21**)

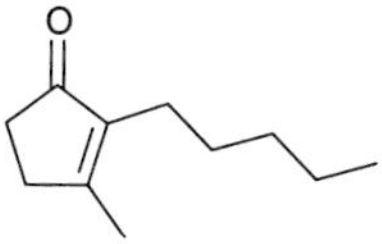

(**22**)
Dihydrojasmone

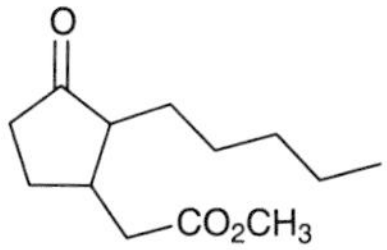

(**23**)
Methyl dihydrojasmonate

However, for lily of the valley this approach has not been successful. Although several groups of workers have studied the composition of the essential oil and concrete of the flowers and stalk, none have identified a component that possesses the true characteristic odour of the flower (Boelens and Wobben, 1980). The closest match is farnesol (3,7,11-trimethyldodeca-2,6,10-trien-1-ol), which has a very mild and delicate sweet-oily odour with a floral fresh-green note that is reminiscent of certain notes from muguet (Arctander, 1969). More recently, by the use of headspace analysis (Chapter 12), dihydrofarnesol has been identified as a fragrance component of lily of the valley (Surburg *et al.*, 1993; Brunke *et al.*, 1996). According to Pelzer *et al.* (1993), dihydrofarnesol is an outstanding lily of the valley fragrance, although its qualities are not apparent on a smelling strip. Its low vapour pressure means that it develops fully only in aerosol form.

Serendipity

Since the blossom oil of lily of the valley is not commercially available, the perfumers have to rely on synthetic substitutes, such as hydroxycitronellal (1), Lilial® (3) and Bourgeonal® (6) to create their muguet fragrances. The oldest lily of the valley odorants, hydroxycitronellal and cyclamen aldehyde (2) (Winthrop Chem. Corp., 1929), were discovered by chance. Serendipity still continues to play an important role. Anselmi *et al.* (1992) synthesized and organoleptically screened a series of 38 tetrahydropyranyl ethers. This class of compounds was chosen because of their ease of synthesis and purification, and because of their stability in alkaline media, conditions under which aldehydes tend to undergo aldol condensations. Two compounds (24 and 25) were described as having a white, floral odour reminiscent of hydroxycitro-

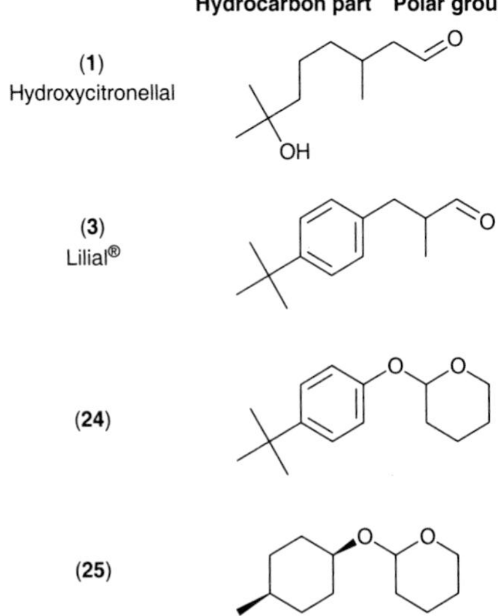

Figure 15.2 *Comparison between the structures of the tetrahydropyranyl ethers of 4-*t*-butylphenol and* cis*-4-methylcyclohexanol, and those of hydroxycitronellal and Lilial®. The structures have been drawn with similar orientation of the polar group to better visualize similarities in size and shape of the hydrocarbon region*

nellal and Lilial®. It was postulated that these four molecules were similar in odour because they could adopt very similar conformations (Figure 15.2) when interacting with olfactory receptors (see later). In these conformations, the size and shape of the hydrocarbon part and the position of the functional group are similar. In the case of hydroxycitronellal, it was suggested that the main role of the hydroxyl group is to provide the necessary bulk at that end of the molecule and that it is not strictly essential as a functional group.

Synthesis of Analogues

Once a compound with the desired odour properties has been discovered it is normal practice to prepare a range of analogues. The reasons for this are two-fold: firstly, materials with even better odour characteristics than the parent compound may be identified (*i.e.* lead optimization) and, secondly, information about the structural requirements for the desired odour characteristics are acquired. The latter is of particular importance when filing a patent.

Following the discovery of the floral tetrahydropyranyl ethers (24) and (25), Anselmi's group synthesized a much wider range of analogues to gain further insight into the relationship between chemical structure and the floral odour of alkoxycycloethers. They investigated the effect of the size of the alkyl substituent, the effect of the position of the alkyl group (limited to a methyl) on the cyclohexane ring, the replacement of the 2-tetrahydropyranyl group by the smaller 2-tetrahydrofuranyl ring (Anselmi *et al*., 1993), and replacement of the oxygen atoms with either sulfur (Anselmi *et al*., 1994a) or carbon (Anselmi *et al*., 1994b).

They found that the position of the methyl group on the cyclohexane ring did not affect the odour appreciably, with all of the derivatives having some floral character. However, increasing the size of the substituents in position 4 drastically reduced the odour intensity. Decreasing the ring size of the cycloether group from C6 to C5 did not affect, on average, the odour of the corresponding derivatives, whereas replacement of either one or both oxygens by sulfur resulted in the loss of the floral odour. The sulfur analogues were generally green and fruity, accompanied in some cases by the unpleasant notes often associated with sulfurous materials. The analogues containing only one oxygen atom, the other being replaced with carbon, were also green and fruity with no floral character.

Another example of analogue synthesis dates back to the 1960s. As mentioned above, the first phenylpropanal to be discovered with a white, floral odour was cyclamen aldehyde, 3-(4-isopropylphenyl)-2-methylpropanal (2). As the name implies, its odour is reminiscent of wild cyclamen. However, at low concentrations the odour possesses a hydroxycitronellal-like note. Consequently, it was common practice in the 1960s to replace hydroxycitronellal, which is unstable in alkaline media and irritating to the skin, with cyclamen aldehyde. However, from an odour point of view the substitution was not entirely satisfactory. The discovery of the closely related *t*-butyl analogues (3 and 6) resulted in a certain improvement (Givaudan Corp., 1959; Boelens and Wobben, 1980). Both of these compounds have a more pronounced muguet floral odour.

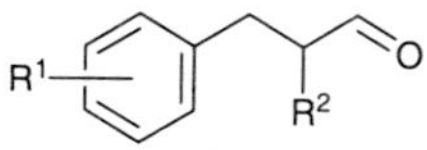

(26)

Berends and van der Linde (1967) synthesized and organoleptically-assessed 24 analogues of cyclamen aldehyde (26). They varied the nature of the alkyl substituent in the *para* position, altered the position of the alkyl group and replaced one of the α-hydrogen atoms with small alkyl groups. This enabled them to make the following qualitative conclusions about the effect of structure on the odour properties of substituted phenylpropanals:

—The cyclamen character is most clearly discernible in compounds with a branched alkyl group in the *para* or *meta* position, where the branch is present at the C atom linked to the phenyl ring [*e.g.* $R^1 = CH(CH_3)_2$].
—Where the branch is at a C atom further removed from the nucleus, the cyclamen character diminishes and the odour becomes more muguet [*e.g.* $R^1 = —CH_2C(CH_3)_3$].
—The introduction of an alkyl substituent at the alpha position in relation to the aldehyde group causes a decrease in odour intensity.
—*Meta* and *para* isomers have similar odour characters and intensity. The *ortho* isomer has a weaker odour of a different type.

STRUCTURE–ACTIVITY RELATIONSHIPS

A set of empirical rules, such as the one above, is the simplest way of describing a SAR. The underlying principle behind any SAR is that the molecular structure of an organic compound determines the properties of that compound, whether it be chemical reactivity, physical properties or biological activity. Thus, a set of compounds which exhibit the same type of activity and which are detected by the same mechanism must have one or more structural features in common. If such a relationship is found, it can be used to predict the activity of compounds that have not yet been made, and as such is a very useful tool in the design of new, potentially active compounds, whether they be pharmacological agents, herbicides, pesticides or chemical communicants (such as odorants and flavorants). Other applications include the investigation of toxicity and the prediction of environmental risk.

The Q is put into QSAR by describing the structure of a compound in a quantitative way, the simplest examples of quantitative descriptors being the mass of the compound or the number of atoms present. When the compound is described using physical, as opposed to structural, properties the relationship becomes a PAR. Correlations of this type have been used in the perfumery industry to describe and predict the substantivity and retention of fragrance ingredients; that is, the ability of a compound to stick to and remain bound to surfaces such as hair, skin or cloth (see Chapter 11 for more details).

The current rapid development of more sophisticated SAR techniques, particularly in the field of drug design, is having a major impact on analogous research in the field of olfaction. The result is a prolific number of structure–odour relationship publications. Advances in theoretical and computational chemistry, coupled with the introduction of computer graphics, have made conformational analysis easier and allowed the visualization of molecules in 3D. This has led to an increase in the number of postulated 'osmophores', which are usually expressed as distance criteria between key structural fragments. Simultaneously, more powerful computers have allowed the rapid calculation of a wide range of molecular descriptors and provided improved data handling. This is reflected in the publication of odour-discriminating models derived using statistical techniques (such as pattern recognition) and, more recently, neural networks. Three SAR techniques that have been applied to the study of odour are described below: the Hansch analysis, the osmophore approach and pattern recognition. However, for a more detailed account on structure–odour relationships, see the recent review of Rossiter (1996a).

The Hansch Approach

The Hansch approach is one of the most widely used methods for analysing structure–activity relationships when quantitative data are available. It is named after the founder of modern QSAR, Corwin Hansch (Hansch and Fujita, 1964), who suggested that the biological activity of a molecule was a function of its electronic, steric and hydrophobic properties; the last most often being represented by the partition coefficient (P) between water and octanol (equation 2).

$$\log(\text{activity}) = f(\text{electronic}) + f(\text{steric}) + f(\log P) \qquad (2)$$

There are two ways of developing a QSAR. The first approach is to set up a model and then test it. So, for example, if we are considering odour intensity we might expect volatility to be important, because the molecules have to become airborne to reach our noses. Log P might also be a relevant parameter because the molecules have to pass through a mucus layer in order to reach the olfactory cilia membrane where the protein receptors are located. When it comes to interaction with the receptor, one might expect hydrogen bonding, electrostatic interactions and shape to be important and would thus use parameters that describe these properties. The disadvantage of this approach is that you are not searching for the unexpected, but only testing your own prejudices. The second approach is to use as many different molecular descriptors as possible and then seek correlations. The parameters can be taken singly or in combinations and may include power terms (*e.g.* x^2, x^3). The main disadvantages with this approach are that it is difficult to limit sensibly the number of parameters, it is difficult to interpret the results, and spurious correlations can be found. To feel confident about the results of a Hansch analysis, and to support these results with statistics, the study set must contain at least five well-chosen compounds for every explanatory variable. For example, an equation, such as equation (2), which has three terms on the right-hand side, should be derived from 15 or more compounds.

A set of well-chosen compounds is one that contains sufficient structural variation to permit the uncovering of a relationship between molecular structure and the activity of interest. It is also important that the series of compounds exhibits varied levels of potency. If a SAR is formulated for a set of very similar compounds, both in terms of activity and structure, the correlation is likely to be very high, but its ability to predict the activity of materials outside this range is likely to be very poor.

Odour can be quantified in one of four ways:

- —As a threshold concentration, *e.g.* the minimum concentration of the compound in the air detected by the nose.
- —As a concentration required to produce the same odour strength as a given dilution of a standard compound.
- —As odour-intensity ratings, which could be as simple as assigning 0 to an odourless compound, 1 to a compound that has a weak odour, 2 to a compound of moderate intensity, and 3 to a very strong-smelling compound.
- —As odour similarity ratings, where the odour quality of one odour type is rated against that of a standard. For example, if a scale of 0 to 10 is used, a compound which is identical in odour to the standard is given a score of 10 and one which is completely different a score of 0.

The last was used by Boelens and Punter (1978) to quantify the odour quality of 16 muguet-smelling materials. These data were then used to derive equation (3), which related the odour similarity (OS) to molecular weight (MW) and the Kier connectivity index (X^v). The concept of molecular connectivity was introduced by Randic (1975) and further elaborated by Kier and Hall (1976). It involves the calculation of numerical indices which describe the topology of a molecule. The Kier connectivity index used in Boelens's equation is a measure of the extent of branching. The correlation coefficient, r, is a measure of how well the model describes the observed variation in the data set. A value of 1 represents a perfect model.

$$\text{OS muguet} = \text{F(MW)} + \text{F}(X^{v}) + \text{C} \qquad (3)$$
$$n = 16, r = 0.92, s = 0.90$$

Based on this work, Boelens designed 4-[4-(1-hydroxy-1-methylethyl)-cyclohexylidenyl]butanal, (27), which, when made, did indeed have excellent olfactive properties. Thus, the activity of compound (27) had been correctly predicted and Boelens's model further substantiated. However, if the prediction had been incorrect it should not, as is quite often the case, be ignored and classified as 'an exception to the rule'. It is important that the reasons for the poor prediction are investigated. Outliers or compounds that exhibit unique biological activity can often provide vital clues about the structural requirements for that biological activity. Thus, the development of a SAR is an iterative process, with the information acquired from new materials being used to refine the model.

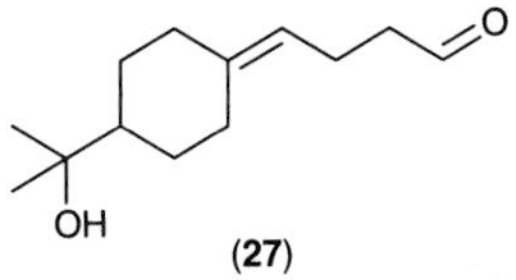

(27)

Boelens has also used this approach to derive QSAR equations for musk, jasmine, fruit and bitter almond odorants (Boelens, 1976; Boelens and Punter, 1978; Boelens *et al.*, 1983). In the case of bitter almond and musk, he concluded that hydrophobic and steric parameters were important. For the jasmine materials, he found that molecular connectivity indices were useful parameters. Molecular connectivity indices were also used by their inventors, Kier *et al.* (1977), to analyse anosmia to fatty acids and the odour similarities of ethereal, floral and benzaldehyde-like odorants. Dearden (1994) also developed a QSAR equation relating the odour similarity of bitter almond odorants to two connectivity indices. Greenberg (1979) found that the odour intensity of a series of homologous compounds was correlated to their hydrophobic properties and not to steric or polar properties, while Rossiter (1996b) found that the fruitiness of aliphatic esters was related to steric hindrance of the ester group and either molecular length or log P.

The Osmophore Approach

The osmophore approach assumes that for a series of active molecules, which are detected by the same mechanism, a common conformation exists in which key atoms or functional groups are placed at certain relative distances from each other. The spatial arrangement is known as a biophore (pharmacophore for drugs, osmophore for odorants) because it can confer a particular activity upon the structure in which it is present. The identification of a biophore model involves the comparison of all the possible conformations of every molecule in the data set. Since this procedure requires no knowledge of the degree of potency, it is well suited to qualitative data; for example, a group of compounds that all smell very similar. The comparison of conformations is made easier by choosing a rigid, active compound as the standard, since this molecule has a limited number of viable conformations that it can adopt. The next step is to take the second-most rigid molecule and look for any conformational arrangements that it has in common with the standard. The procedure is repeated for every compound in the data set until, hopefully, only a handful of common conformational arrangements are left. All of the osmophore models identified are explored.

The models can be tested using inactive molecules. These, in theory, should not be able to satisfy the proposed biophore. However, inactivity can be the result of other factors. For example, a molecule that can fulfil the distance geometry requirements may be inactive because it cannot reach the site of detection. The latter is determined by physicochemical properties such as hydrophobicity, solubility and volatility. The main criticism of the osmophore approach is that it does not take into account these types of properties.

Muguet odorants are not well suited to the osmophore approach because most of them exhibit a high degree of conformational flexibility and thus have thousands of possible conformational arrangements. To overcome this, Pelzer *et al.* (1993) considered only the energy-minimized forms of the 20 best muguet fragrance ingredients. However, it is generally accepted that the active conformation is not necessarily the lowest energy one. Burgen *et al.* (1975) suggested that when a flexible drug molecule binds to a receptor it is probable that a nucleation complex is first formed by association of part of the drug with a sub-site on the receptor. Presumably, a similar thing could happen in olfaction. Pelzer, in his muguet study, found that if the oxygen atom is chosen as a fixed point, the position and substitution pattern of the carbon atoms C-4, C-5 and C-6 play an important role. However, when this model

Osmophore 1

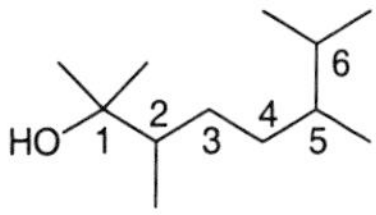

C1–C4 = 3.2 ± 0.3 Å
C1–C5 = 4.0 ± 0.3 Å
C1–C6 = 5.0 ± 0.3 Å
C1–R1 = 4.7 ± 0.4 Å
C1–R2 = 6.0 ± 0.4 Å

Osmophore 2

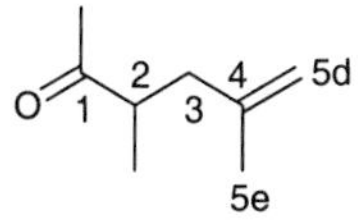

C1–C4 = 3.9 ± 0.3 Å
C1–C5d = 4.4 ± 0.3 Å
C1–C5e = 5.0 ± 0.3 Å

Figure 15.3 *Pelzer's muguet osmophores*

was tested using a further 161 compounds, some of the compounds showed a different substitution pattern. It was concluded that there were two classes of muguet odorants; those that have a hydroxyl group and those with a carbonyl group. The two types were shown to have different odour profiles, with the carbonyl compounds exhibiting, in addition to the muguet aspect, 'lime blossom' notes. The substitution pattern and the distance constraints for the osmophores of these two classes are shown in Figure 15.3.

Pelzers' analysis of 73 alcohols produced the following rules for osmophore 1:

—C-l is substituted by one to three alkyl groups, ideally three, provided that the hydroxyl group is not overshadowed too strongly by steric hindrance.
—C-2, C-5 and to a lesser extent C-6 and C-7 are substituted by a single alkyl group (ideally methyl). Dimethyl substitution generally has a detrimental effect.
—Where a double bond is present, it should preferably be at C-4 or C-6. A double bond between C-3 and C-4 or, to a lesser extent, between C-2 and C-3, generally has a negative influence on the odour.

For osmophore 2 it was found that:

—An aldehyde is always better than a ketone function.
—C-2 should be substituted by one or two methyl groups, preferably one.
—C-4 should be alkyl-substituted; a double bond at C-4 is particularly advantageous, and may also be part of an aromatic system.

The use of these models led to the synthesis of a series of interesting new fragrances, the most notable of which was Mugetanol® (28).

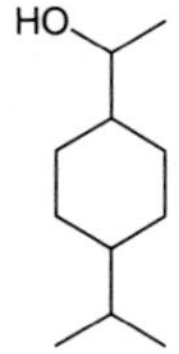

HO

(28)
Mugetanol® (Haarmann & Reimer)

Conformational analysis and the identification of osmophores is best suited to sets of compounds with a limited amount of conformational flexibility. For these reasons, this approach has also been applied to ambergris (Bersuker *et al.*, 1985; Dimoglo *et al.*, 1995), musk (Bersuker *et al.*, 1991; Chastrette and Zakarya, 1988) and sandalwood (Chastrette *et al.*, 1990; Buchbauer *et al.*, 1994; Aulchenko *et al.*, 1995) odorants. One of the ambergris osmophores from the recent SAR study of Bajgrowicz and Broger (1995) has been successfully used in the design of structurally novel ambergris odorants (Fráter *et al.*, 1998; Bajgrowicz, 1997).

Pattern Recognition

Pattern recognition deals with the problem of dividing compounds into two or more classes. It is thus well suited to analysing qualitative SARs for which the biological activity data are of a descriptive nature (*e.g.* odour descriptors) or of ranges such as strong, moderate, weak or inactive. Other characteristics of the technique are that it can handle large numbers of compounds of substantial structural diversity. To obtain meaningful discrimination between the classes it is important that the inactive molecules are as close as possible in structure to the active ones.

One of the software systems available for pattern recognition studies is ADAPT (automated data analysis using pattern recognition techniques). The structure of each member of the data set is represented by molecular descriptors. These numerical indices, which encode information about the molecule, fall into four classes: topological, geometrical, electronic and physicochemical. The data are analysed using pattern recognition techniques to develop a classifier which can discriminate between the classes of data.

ADAPT has been developed and used by Jurs in a wide range of SAR applications. In the field of olfaction these include the correlation of odour intensities for 58 structurally and organoleptically diverse odorants (Edwards and Jurs, 1989), and the investigation of the relationship between molecular structure and musk odour (Ham and Jurs, 1985; Narvaez *et al.*, 1985; Jurs and Ham, 1977). To date, no one has used pattern recognition techniques in the study of muguet odorants.

A range of other statistical techniques can be used in the formulation of a classification model. Since a detailed description of these is outside the scope of this chapter, those which have been used in the study of odour are listed below:

—Principal component analysis: bitter almond (Zakarya *et al.*, 1993), fruit odour (Rossiter, 1996b).
—Linear discriminant analysis: bitter almond (Zakarya *et al.*, 1993), camphor (Chastrette and Eminet, 1983), musk (Klopman and Ptchelintsev, 1992; Yoshii *et al.*, 1991), sandalwood (Chastrette *et al.*, 1990).
—Neural networks: musk (Chastrette *et al.*, 1993; Chastrette *et al.*, 1994; Jain *et al.*, 1994).

LIGAND–OLFACTORY RECEPTOR MODELLING

Ingredient design, in theory, may be aided by the computerized molecular modelling of ligand–receptor interactions. To date, there are only three publications of this type in the area of olfaction (Singer and Shepherd, 1994a, 1994b; Bajgrowicz and Broger, 1995). Two involve the identification of hypothetical muguet binding sites in the rat receptor protein (OR5). Singer and Shepherd (1994a) carried out docking experiments with Lyral® (4). The results point to a potential binding pocket which is made from protein residues distributed in helices 3 through to 7. One year later Bajgrowicz and Broger (1995) reported their work using Lilial® (3) as the ligand. The putative binding site obtained from this experiment was formed from eleven amino acids of helices 3, 4, 6 and 7. The results from the ligand–receptor modelling support the suggestion by Buck and Axel (1991), that helices 3, 4 and 5, which exhibit wide protein sequence diversity from one receptor to another, are involved in odorant binding.

Buck and Axel were the first group of workers to clone a range of proteins from rat olfactory tissue. Other workers have extended the cloning to mice, humans, dogs and fish, producing hundreds of candidate molecules that could be olfactory receptor proteins. These proteins belong to the superfamily of seven transmembrane domain proteins, so named because each one is stitched seven times through the cell membrane. They are still poorly characterized and their binding affinity for odorants has yet to be demonstrated, although there is suggestive evidence that a number of these receptors respond positively to odour molecules (Raming *et al.*, 1993; Kiefer *et al.*, 1996; Sengupta *et al.*, 1996; Firestein *et al.*, 1998). For example, Raming *et al.* (1993) expressed the rat receptor protein (OR5) in the baculovirus-Sf9 cell, which when stimulated with a mixture of Lyral® and Lilial® showed a two-fold increase in the level of inositol triphosphate (IP_3). The latter is one of the secondary messengers believed to be involved in odour transduction (Breer, 1993). It was the positive results from Raming's work which inspired the aforementioned ligand–receptor modelling experiments. However, the modelling of this important class of proteins is highly speculative because of insufficient knowledge of the olfactory mechanisms, and the assumptions which have to be made regarding the tertiary structure of the protein. Thus, models of this type are incorrect in many parts, but nonetheless are very useful in posing experiments to probe structure–odour relationships further. It will be some time before computers can translate a gene sequence into a 3D model of the receptor and screen thousands of odour molecules to find a few that optimally activate it.

SUMMARY

At the beginning of this chapter it was explained why there is a continuing need to extend the perfumer's palette. The discovery of new, patentable fragrance ingredients, which are either cheaper, more stable and more readily biodegradable than existing aroma chemicals or which have a unique odour or additional functionality, gives Quest's perfumers a competitive edge. Serendipity and the analysis of natural products still play an important role. However, as progress in the biological sciences leads to an increased understanding of the mechanism of olfaction and as more sophisticated SAR tools become developed, the search for correlations between structure and odour should become easier. This challenge, coupled with the potential predictive ability of this approach, will entice chemists and molecular modellers to continue research in this area. The future of fragrance ingredient design promises to be both challenging and exciting.

ACKNOWLEDGEMENTS

Ian Payne, Quest International, for organoleptical stability data of muguet ingredients. Steven Rowlands, Quest International, for analysis of the breakdown products of Lilial® in antiperspirant base.

Andy Roche, Reheis, for information on the partial hydrolysis of aluminium chlorhydrate.

REFERENCES

C. Anselmi, M. Centini, M. Mariani, A. Sega and P. Pelosi, *J. Agric. Food Chem.*, 1992, **40**, 853.

C. Anselmi, M. Centini, M. Mariani, A. Sega and P. Pelosi, *J. Agric. Food Chem.*, 1993, **41**, 781.

C. Anselmi, M. Centini, M. Mariani, A. Sega and P. Pelosi, *J. Agric. Food Chem.*, 1994a, **42**, 1332.

C. Anselmi, M. Centini, M. Mariani, E. Napolitano, A. Sega and P. Pelosi. *J. Agric. Food Chem.*, 1994b, **42**, 2876.

S. Arctander, in *Perfume and Flavor Chemicals*, published by the author, Montclair, NJ, 1969.

I. S. Aulchenko, A. A. Beda, A. S. Dimoglo, M. Yu. Gorbachov, L. A. Kheifits and N. M. Shvets, *New J. Chem.*, 1995, **19**, 149.

J. A. Bajgrowicz and C. Broger, in *Flavours, Fragrances, and Essential Oils. Proceedings of the 13th International Congress of Flavours, Fragrances and Essential Oils*, K. H. C. Baser (ed.), AREP Publ., Istanbul, 1995, vol. 3, p. 1.

J. A. Bajgrowicz, EP 0.761.664, 1997.

W. Berends and L. M. van der Linde, *Perfum. Essent. Oil Rec.*, 1967, 372.

I. B. Bersuker, A. S. Dimoglo, M. Yu. Gorbachov, M. N. Koltsa and P. F. Vlad, *New J. Chem*, 1985, **9**(3), 211.

I. B. Bersuker, A. S. Dimoglo, M. Yu. Gorbachov and P. F. Vlad, *New J. Chem.*, 1991, **15**, 307.

H. Boelens, in *Structure–Activity Relationships in Chemoreception*, G. Benz (ed.), Information Retrieval Ltd, London, 1976, p. 197.

H. Boelens, H. G. Haring and H. J. Takken, *Chem. Ind.*, 1983, 26.

H. Boelens and P. Punter, in *Third ECRO Congress*, European Chemoreception Research Organization, Italy, 1978, 1.

H. Boelens and H. J. Wobben, *Perfum. Flavorist*, 1980, **5**(6), 1.

H. Breer, in *The Molecular Basis of Smell and Transduction*, Wiley, Chichester, 1993, 97.

H. Breer, K. Raming and J. Krieger, *J. Biochim. Biophys. Acta*, 1994, **1224**, 277.

E. J. Brunke, F. J. Hammerschmidt, F. Rittler and G. Schmaus, *SOFW J.*, 1996, **122**, 593.

G. Buchbauer, A. Hillisch, K. Mraz and P. Wolschann, *Helv. Chim. Acta*, 1994, **77**, 2286.

L. Buck and R. Axel, *Cell*, 1991, **65**, 175.

A. S. V. Burgen, G. C. K. Roberts and J. Feeny, *Nature*, 1975, **253**, 753.

M. Chastrette and B. P. Eminet, *Chem. Senses*, 1983, **7**, 293.

M. Chastrette and D. Zakarya, *C. R. Acad. Sci. Paris, série II*, 1988, **307**, 1185.

M. Chastrette, D. Zakarya and C. Pierre, *Eur. J. Med. Chem.*, 1990, **25**, 433.

M. Chastrette, J. Y. De Saint Laumer and J. F. Peyraud, *SAR QSAR Environ. Res.*, 1993, **1**(2–3), 221.

M. Chastrette, D. Zakarya and J. F. Peyraud, *Eur. J. Med. Chem.*, 1994, **29**, 343.

J. C. Dearden, *Food Quality Preference*, 1994, **5**, 81.

A. S. Dimoglo, P. F. Vlad, N. M. Shvets, M. N. Coltsa, Y. Güzel, M. Saraçoğlu, E. Saripinar and Ş. Patat, *New J. Chem.*, 1995, **19**, 1.
P. A. Edwards and P. C. Jurs, *Chem. Senses*, 1989, **14**(2), 281.
S. Firestein, H. Zhao, L. Ivic, J. M. Otaki, M. Hashimoto and K. Mikoshiba, *Science*, 1998, **279**, 237.
G. Fráter, J. A. Bajgrowicz and P. Kraft, *Tetrahedron*, 1998, **54**, 7633.
The Givaudan Corp., US 2.875.131, 1959.
The Givaudan Corp., US 2.976.321, 1961.
M. J. Greenberg, *J. Agric. Food Chem.*, 1979, **27**, 347.
C. L. Ham and P. C. Jurs, *Chem. Senses*, 1985, **10**, 491.
C. Hansch and T. Fujita, *J. Am. Chem. Soc.*, 1964, **86**, 1616.
A. N. Jain, T. G. Dietterich, R. H. Lathrop, D. Chapman, R. E. Critchlow Jr., B. E. Bauer, T. A. Webster and T. J. Lozano-Perez, *J. Comp. Aided Design*, 1994, **8**, 635.
P. C. Jurs and C. L. Ham, *J. Agric. Food Chem.*, 1977, **25**(5), 1158.
H. Kiefer, J. Krieger, J. D. Olszewski, G. von Heijne, G. D. Prestwich and H. Breer, *Biochemistry*, 1996, **35**, 16077.
L. B. Kier and L. H. Hall, in *Molecular Connectivity in Chemistry and Drug Research*, G. de Stevens (ed.), Academic Press, New York, 1976.
L. B. Kier, T. DiPaolo and L. H. Hall, *J. Theor. Biol.*, 1977, **67**, 585.
G. Klopman and D. Ptchelintsev, *J. Agric. Food Chem.*, 1992, **40**, 2244.
P. Laffort, in *Odors and Deodorization in the Environment*, G. Martin and P. Laffort (eds), VCH Publishers, New York, 1994, 143.
J. N. Narvaez, B. K. Lavine and P. C. Jurs, *Chem. Senses*, 1985, **10**, 145.
P. Pelosi, *J. Neurobiol.*, 1996, **30**(1), 3.
R. Pelzer, U. Harder, A. Krempel, H. Sommer, H. Surburg and P. Hoever, in *Recent Developments of Flavour and Fragrance Chemistry: Proceedings of the 3rd International Haarmann & Reimer Symposium*, R. Hopp and K. Mori (eds), VCH Publishers, Weinheim, 1993, 29.
K. Raming, J. Kreiger, J. Strotmann, I. Boekhoff, S. Kubik, C. Baumstark and H. Breer, *Nature*, 1993, **361**, 353.
M. Randic, *J. Am. Chem. Soc.*, 1975, **97**, 6609.
G. V. Ronnett, in *Handbook of Olfaction and Gustation*, R. L. Doty (ed.), Marcel Dekker Inc., New York, 1995, 127.
K. J. Rossiter, *Chem. Rev.*, 1996a, **96**(8), 3201.
K. J. Rossiter, *Perfum. Flavorist*, 1996b, **21**(2), 33.
C. S. Sell, *Chem. Britain*, 1997, **33**(3), 39.
P. Sengupta, J. H. Chou and C. I. Bargmann, *Cell*, 1996, **84**, 899.
M. S. Singer and G. M. Shepherd, *NeuroReport*, 1994a, **5**, 1297.
M. S. Singer and G. M. Shepherd, in *Abstracts of the 16th Annual Meeting of the Association for Chemoreception Sciences*, Sarasota, 1994b, 207.
H. Surburg, M. Guentert and H. Harder, in *Recent Developments of Flavour and Fragrance Chemistry: Proceedings of the 3rd International Haarman & Reimer Symposium*, R. Hopp and K. Mori (eds.), VCH Publishers, Weinheim, 1993, p. 29.
Winthrop Chem. Corp., US 1.844.013, 1929.
F. Yoshii, Q. Liu, S. Hirono and I. Moriguchi, *Chem. Senses*, 1991, **14**(4), 319.
D. Zakarya, M. Yahiaoui and A. Fkih-Tetouani, *J. Phys. Org. Chem.*, 1993, **6**, 627.

Chapter 16

The Brief Submission

DAVID PYBUS AND LES SMALL

MARKETING PLATFORM

While the brief from Business Scents Ltd for 'Eve' allowed a deal of creative flexibility, there are certain key points which help focus our thoughts for a marketing platform. These are:

—The basic structure and composition of the fragrance itself. A story needs to complement and underscore the make-up of the perfume.
—The core-group target of global females between 16 and 35 years old.
—The fact that a range extension was likely.
—Novelty in approach was sought, and an idea which suited the new age of not only a new century, but a new Millennium too.

Whilst it is unusual for a client to ask for too much direction on the marketing side, a more common event is for a fragrance house to proactively approach a client with a full-blown idea complete with fragrance and story boards, much along the lines of an advertising agency. Many new products on the market today have had their genesis in this manner. What started out as an 'Eve' brief from the client metamorphosed into our presentation of *Djinni*.

DJINNI CONCEPT

For Business Scents Ltd we came up with the concept of *Djinni*®. Djinni is an Arabic term for 'Genie', and immediately evocative of the

Middle East, romance, adventure and fable. A play on the word 'Jeannie', the girl's name, it could use a similar platform to that of the first lifestyle fragrance of the seventies called *Charlie*®, which had a very strong and successful launch, and was aimed at a similar audience to that of 'Djinni'.

—*Djinni*® is also the spirit in the bottle, waiting to be unchained or released; another possible theme (like a caged animal being set free).
—For a nostalgic approach, the use of the old song 'I dream of Jeannie' could be made.
—The recent and much acclaimed Disney cartoon 'Aladdin' means that the symbol of a Genie/Djinni is very strong, known and recognized the world over.
—The two syllable word 'Djinni' is easy to say and pronounce.
—A novel approach would be to launch it through an encapsulation strip in magazines, as a tester (see back inside page), where rubbing a lamp releases the spirit (the perfume).
—A play on words with Djinni (or Jinn) and Gin as the 'spirit of the bottle' could be used as a fun play on words in advertising or public relations.
—Further strong symbolism could be evoked if the glass container for the fragrance is in the form of a lamp; again, a novel approach.
—Djinni supports some of the 'oriental/tropical' composition of the fragrance, and has a name which is both familiar yet different; a name that is exotic.
—Finally, a useful environmental message could be pursued in offering 'new lamps for old'; that is, in offering a potential refill service. The strong eco-message here would have to be balanced against what kind of image *Djinni* wished to project.

FRAGRANCE DESCRIPTION OF 'DJINNI'

Head and Top Notes: Fruity–Citrus

Composed of apple, grapefruit and a melange of tropical fruits, such as papaya, mango and passion fruit, with sweet, herbaceous camomile, which also has fresh, fruity undertones. This volatile mix signals a heady and early release of the Djinni from the bottle.

Middle and Heart Notes: Watery Muguet (Lily of the Valley)

The heart consists of a floral bouquet majoring on muguet (lily of the valley) and violet leaves, with a background of spicy carnation, narcotic jasmine and the sweet florality of broom.

Base and Foundation Notes: Powdery Amber

A sweet, oriental base is provided by animalic, synthetic musk, sensual amber, warm sandalwood and powdery vanilla.

Overall Composition

The complete composition evokes the mythical being (the prisoner of the bottle), the spirit of the user. The creation has taken into account both the framework of the customer's brief and the chemistry of the fine-fragrance and the toiletry range based on it. With adaption, it is robust enough for the chemical environment of soaps, antiperspirants, shampoos and shower gels.

It should have global appeal, particularly in the key markets of Europe, the USA and Asia Pacific, based on both current and historical research. The 'smell direction' is relatively novel, whilst not being too revolutionary, as consumers tend to be relatively conservative and seek a combination of novelty with familiarity.

Whilst another floral type, thus complementing *Ninevah*, the first fragrance launch of 1994 by Business Scents Ltd, *Djinni* is a much lighter composition, fruitier (with some novelty of scope) and more in line with the trend towards 'transparency'. It is thus likely to be more acceptable to the audience of wide age range indicated by the client, than the heavier floriental type represented by *Ninevah*.

Epilogue

Proactivity in the fragrance industry makes us seek out new briefs and forge ahead with new chemistry that will aid in winning future business using techniques and aromas that could well not exist today.

To chemists of all persuasions the challenge of fragrances is an exciting and creative one, which relates directly to consumer products that we all use, and which can delight people, enhancing their lives and bringing a mixture of passion and practicality into the world.

We have taken up the magic wand handed down from alchemists of ancient times, and renowned chemists of the recent past, and are ready, in turn to refine and pass on that knowledge.

Perfume is a good example of an emergent phenomenon, something which 'emerges' when there is a deal of complexity in the system, or a sufficient quantity of molecules to demonstrate a specific characteristic. When a bottle of fragrance is unstoppered and left open in an enclosed space, after a while the volatile mix of chemicals will diffuse throughout the air and is detectable in the room. The perfume industry 'reverses entropy' in that a cornucopia of Nature's bounty is brought together in a bottle in a creative, structured manner for a finite moment of time.

Unleashed, however, perfume molecules stream into the air, mixing and rebounding with air molecules in an irreversible process, since it would be more than difficult to put the aroma back in the bottle once the scent molecules are released, and we can be fairly sure that the random molecular motions of the perfume are unlikely, even over a great length of time, to reverse the process and go back into the bottle of their own accord.

Thus, arguably irreversible, the release of perfume from a bottle, as an emergent phenomenon, has an 'arrow of time', as observed when the molecular mix is taken as a whole. But any one molecule in a fragrance, consisting of any of the myriad aromas used in its creation, has no temporal directionality, as it simply reacts to random collisions with air molecules. The Djinni, once released, cannot be put back into the bottle.

We hope you enjoyed the chemistry of our scent trail, and invite you finally to let the Djinni free from the lamp on the inside back page.

Appendix I

ALCOHOL

Chemists place the solvent in the first rank among all assistants and boast that, with its aid, they can perform all the miracles of their art.

Hermann Boerhavve (1668–1738)

It was mentioned earlier in the book that it takes three ingredients to make a fine-fragrance. The aroma chemical mix, the fine cut-glass bottle to capture them and the 'elixir' ethanol. The specification for 99.7% pure ethyl alcohol*, as used in the mass fragrance market where cost-effectiveness and consistency of quality is assured, is detailed in Table A1. The product contains traces of Bitrex to make it undrinkable, and thus not dutiable as such.

With a faint 'smell' of alchemical heritage, the European fine fragrance industry, by contrast, uses in the main alcohol derived from molasses. Being natural, this alcohol has the combination of an odour, a slight inconsistency of quality and a living essence of its own, which is highly valued in an industry which places uniqueness and character on a high pedestal.

Table A2 illustrates essentially what the perfume–ethanol–water mix is in a fragrance bottle, denoted by its description. Figure A1 traces the historical growth of alchemical tradition from the five main centres of civilization through 5000 years of history to the present.

Finally, all three magical links are together. Alcohol, aroma chemicals and silica. The Djinni in a bottle that encapsulates tens of thousands of years of history.

*Courtesy of BP Chemicals.

Table A1 *Ethanol DEB grades (denatured Ethanol 'B'): Sales Specification*

Parameter	*Units*	*Value*[a] *DEB 96*	*DEB 100*	*Test method*
Alcohol content	% vol. at 20 °C	96.1 max. 95.7 min.	100.0 max. 99.7 min.	OIML
Water content	% mass	6.6 max.	0.5 max.	BS 2511:1970
Acidity	% mass as acetic acid (fixed)	0.003 max.	0.003 max.	BP Chemicals method
Total carbonyls	% mass as acetaldehyde	0.02 max.	0.02 max.	BS 6392/2 ISO 1388/2
Appearance		Clear, and free from suspended matter	Clear, and free from suspended matter	BP Chemicals method
Colour	Hazen	10 max.	10 max.	BP Chemicals method
Miscibility with water		Complete	Complete	BS 6392/5 ISO 1388/6
Residue on evaporation	% mass	0.005 max.	0.005 max.	BS 4524:1983 ISO 759:1981

[a] BP Chemicals Ethanol DEB grades comply with the following regulations/recommendations:
Methylated Spirits Regulations S12009, 1987.
CTPA recommendation for use in Cosmetics and Toiletries.

Table A2 *What's in the bottle?*[a]

Nature of branded fragrance	*% fragrance concentrate in ethanol*	*Alcohol specification (% in water)*
Extrait or parfurn	15–30	96–99
Parfum de toilette	8–15	85–90
Eau de parfum		
Esprit de parfum		
Eau de toilette/toilet water	4–15	80–90
Eau de cologne	3–5	60–80
After shave	2–8	50–70
Splash cologne	2–3	50–70

[a] For example, a parfum could be 30% concentrated fragrance in 70% of a 99% pure alcohol, whilst a splash cologne may contain 3% concentrate in a 70% ethanol and 27% pure water mix.

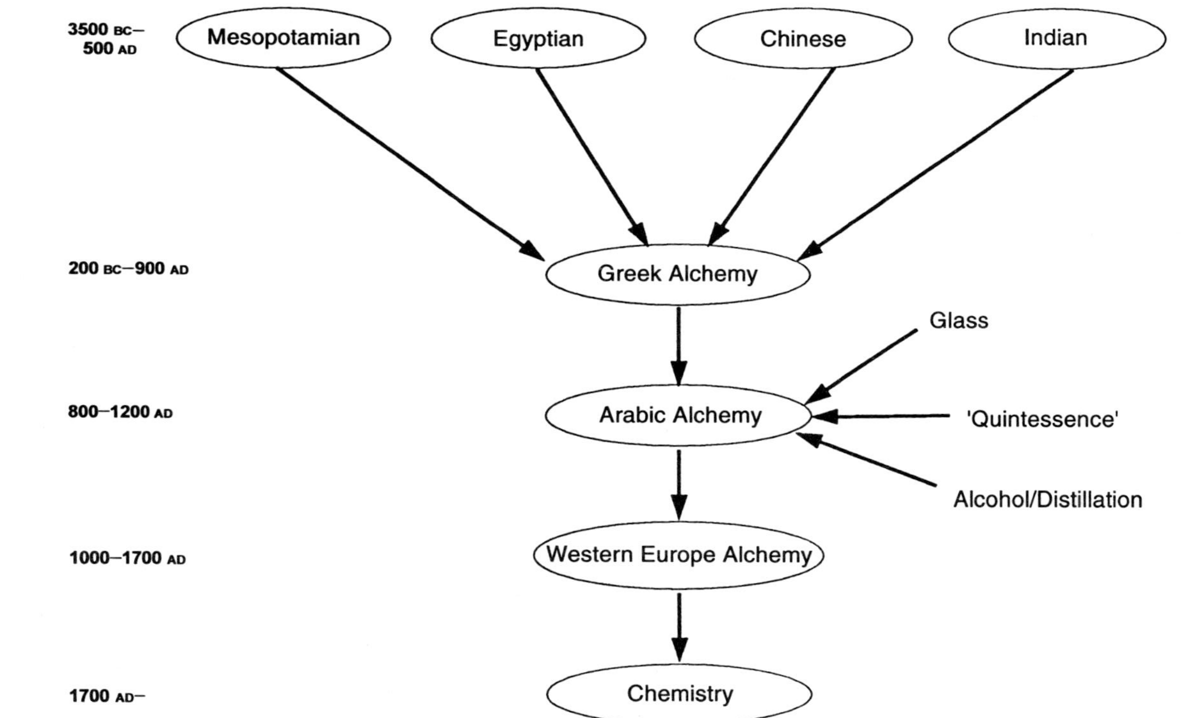

Figure A.1 *The flow of alchemy*

Appendix II

USEFUL ADDRESSES

American Cosmetic, Toiletry and Fragrance Association, 1101 17th St. NW Suite 300, Washington DC 20036, USA

American Oil Chemists Society, PO Box 3489 Champaigne, IL 61826-3489, USA

American Society of Perfumers, PO Box 394, Bergenfield, New Jersey 07621, USA

Aroma Research Institute of America, PO Box 282, Santa Fe, NM87504, USA

Association des Ingenieurs et Techniciens de la Parfumerie (AITP) (also Sydicat des Fabricants des Produits Aromatiques), BP 16-48 Avenue Riou Blanquet, 06130 Grasse, France

BP Chemicals Ltd. (Ethanol), Belgrave House, 76 Buckingham Palace Road, London SW1 0SU, UK

British Fragrance Association, 6 Catherine Street, London WC2B 5JJ, UK

British Society of Perfumers, Glebe Farmhouse, Mears Ashby Road, Wilby, Wellingborough, Northamptonshire NN8 2UQ, UK

COLIPA (European Cosmetic, Toiletry and Perfumery Association), Rue De La Loi, 223/2-B-1040 Brussels, Belgium

Comite Francais du Parfum (CEP), 57 Avenue de Villiers, 75017 Paris, France

Dutch Cosmetic Society, Gebouw Trindenborch, Catharijnesingel 53, 3511 Utrecht, Netherlands

Federation Francais de l'Industrie des Produits de Parfumerie, de Beaute et de Toilette, 8 Place de General Catroux, 75017 Paris, France

Federation Francais de Syndicats de l'Aromatique (FEDAROM), 89 Rue du Faubourg St. Honore, 75008 Paris, France

French Cosmetic Society, 17 Rue Carl-Hebert-92400, Courbevoie, France

German Society of Perfumers, Heinrich-Heine Strasse 25, D-7100 Heilbronn, RFA

Institut Superieur Internacional du Parfum, de la Cosmetique et de l'Aromatique Alimentaire, 36 Rue du Parc de Clagny, 78000 Versailles, France

International Federation of Cosmetic Societies, Delaporte House, 57 Guildford Street, Luton, Bedfordshire LU1 2NL, UK

International Fragrance Association (IFRA), 8 Rue Charles Humbert, CH 1025 Geneva, Switzerland

Italian Federation of Cosmetic, Toiletry and Perfumery Industries (FENA-PRO), Corso Venezia 47, Milan 20121, Italy

Italian National Association of Perfumery and Cosmetic Industries, Via Juvara 9, 20129 Milan, Italy

Japanese Cosmetic Industry Association, 4th Floor Hatsumei Building, 9-14-2 Chom Toranomon, Minatoku, Tokyo, Japan

Lalique (glass flagons for perfumery), 11 Rue Royale, 75008 Paris, France

Monell Chemical Senses Centre, 3500 Market Street, Philadelphia PA 19104-3308, USA

Perfumery Education Centre, University of Plymouth Business School, Drake's Circus, Plymouth, Devon PL4 8AA, UK

Quest International Fragrance Division HQ, Willsborough Road, Ashford, Kent TN24 0LT, UK

Research Institute of Fragrance Materials (RIFM), 375 Sylvan Avenue, Englewood Cliffs, New Jersey 07632, USA

Society of Cosmetic Chemists, GT House, 24/26 Rothesay Road, Luton, Bedfordshire LU1 1QX, UK

Spanish Perfumer Group, Pau Claris, 107 pral.-08009, Barcelona, Spain

The Chartered Institute of Marketing (CIM), Moor Hall, Cookham, Maidenhead, Berkshire SL6 9QH, UK

The Cosmetic, Toiletry and Fragrance Association, Suite 800, 1110 Vermont Avenue, Washington DC 20005, USA

The Cosmetic, Toiletry and Perfume Association (CTPA), 35 Dover Street, London W1X 3RA, UK

The Fragrance Foundation, 145 East 32nd Street, New York, 10016-6002, USA

The International Federation of Essential Oils and Aroma Trades (IFEAT), Federation House, 6 Catherine Street, London WC2B 5JJ, UK

The New School of Perfumery Art, 15 Winding Way, Verona, New Jersey 070440, USA

The Royal Society of Chemistry, Thomas Graham House, Science Park, Milton Road, Cambridge CB4 0WF, UK

Universite de Montpelier Faculty of Science and Technology of Languedoc, Place Eugene Bataillon, 34060 Montpellier, France

Women in Flavor and Fragrance Commerce, PO Box 2154, Teaneck, New Jersey 07666, USA

Bibliography

* Recommend starting point in each section.

GENERAL

T. Curtis and D. G. Williams, *Introduction to Perfumery*, Ellis Horwood, 1995.

*H. V. Daeniker, *Flavours and Fragrances Worldwide*, SRI, 1987.

Gloss, *The H and R Book of Perfume*, 1991.

Gloss, *The H and R Fragrance Guide*, 1991.

*P. M. Muller and D. Lamparsky (eds.), *Perfumes, Art, Science and Technology*, Elsevier, 1991.

G. Ohloff, *Scent and Fragrances*, Springer, 1994.

B. T. Theimer (ed.), *Fragrance Chemistry*, Academic Press, 1982.

HISTORY OF PERFUMERY

C. Baudelaire, *Les Fleurs du Mal*, Harvester, 1986.

BP Chemicals, *Oxygenated Solvents: Cosmetics Science to a Fine Art*, BP Chemicals, 1993.

Brewer's Dictionary of Phrase and Fable, Centenary Edition, Cassell, 1977.

M. Caron and S. Hutin, *The Alchemists*, Barrie and Jenkins, 1974.

Comité Français du Parfum, *Heavenly Scent*, 1995.

B. Dobbs, *The Foundations of Newton's Alchemy*, Cambridge University Press, 1975.

R. Genders, *A History of Scent*, Hamish Hamilton, 1972.

N. Groom, *The Perfume Book*, Chapman and Hall, 1992.

M. Haeffner, *Dictionary of Alchemy*, Aquarian, 1991.

A. J. Hopkins, *Alchemy, Child of Greek Philosophy*, New York Press, 1934.

F. Kennett, Richard Howard translation, *A History of Perfume*, Maynard, 1975.

S. Klossowki de Rola, *The Golden Game*, Thames and Hudson, 1988.

E. Launert, *Scent and Scent Bottles*, Barrie and Jenkins, 1974.
S. Mahdihassan, *Alchemy in the Light*, Janus, 1961.
S. Nefurazi, *The Perfumed Garden*, Spearman, 1963.
R. Norvill, *Language of the Gods*, Ashgrove, 1987.
R. Norvill, *Hermes Unveiled*, Ashgrove, 1986.
Pilkington Glass Museum, *A History of Glass*, Pilkington Glass Museum, 1990.
J. Read, *Prelude to Chemistry*, G. Bell & Son, 1961.
A. H. Rose, *Alcoholic Beverage Production*, Bath University, 1977.
F. S. Taylor, *The Alchemists, Founders of Modern Chemistry*, Schuman, 1949.
J. Trueman, *The Romantic Story of Scent*, Aldus Books, 1975.
A. Walker, *Scent Bottles*, Shire Publications, 1974.
R. Wilhelm, *The Secret of the Golden Flower*, Paul Routledge & Kegan, 1995.

NATURAL PERFUMERY MATERIALS

*S. Arctander, *Perfume and Flavour Materials of Natural Origin*, S. Arctander, 1960.
*E. Gildemeister and Fr. Hoffmann, *Die Ätherischen Öle*, 11 volumes, Akademie-Verlag, 1956.
*E. Gunther, *The Essential Oils*, 6 volumes, D van Nostrand, 1948.
R. K. M. Hay and P. G. Waterman (eds), *Volatile Oil Crops: Their biology, Biochemistry and Production*, Longman, 1993.
B. M. Lawrence, A Review of the World Production of Essential Oils, *Perfum. Flavorist*, 1985, **10**(5), 1.
B. D. Mookherjee and C. J. Mussinian, *Essential Oils*, Allured, 1981.
E. J. Parry, *The Chemistry of Essential Oils and Artificial Perfumes*, Scott, Greenwood and Son, 1921.
Unilever, Essential Oils, Unilever Educational Booklet, Unilever, 1961.
D. G. Williams, *Chemistry of Essential Oils*, Michelle Press, 1996.

SYNTHETIC PERFUMERY MATERIALS

*S. Arctander, *Perfume and Flavour Chemicals*, 2 volumes, S. Arctander, 1969.
K. Bauer and D. Garbe, *Common Fragrance and Flavour Materials*, VCH Publishers, 1985.
K. Bauer, D. Garbe and H. Surburg, *Common Fragrance and Flavour Materials*, 2nd edn, VCH Publishers, 1990.

P. Z. Bedoukian, *Perfumery Synthetics and Isolates*, D van Nostrand, 1951.

*P. Z. Bedoukian, *Perfumery and Flavoring Synthetics*, Elsevier, 1967.

R. Croteau (ed.), *Fragrance and Flavour Substances*, D and PS Verlag, 1980.

W. E. Dorland and J. A. Rogers, *The Fragrance and Flavor Industry*, W. E. Dorland, 1977.

Givaudan-Roure, *The Givaudan-Roure Index*, Givaudan-Roure, 1994.

R. W. James, *Fragrance Technology, Natural and Synthetic Perfumes*, Noyes Data Corporation, 1975.

R. W. Moncrieff, *The Chemistry of Perfumery Materials*, United Trade Press, 1949.

W. A. Poucher, *Perfumes, Cosmetics and Soaps*, Chapman and Hall, 1959.

T. F. West, H. J. Strausz and D. H. R. Barton, *Synthetic Perfumes*, Edward Arnold, 1949.

OLFACTION

*J. F. Amoore, *Molecular Basis of Odour*, Charles C. Thomas, 1970.

R. L. Doty (ed.), *Handbook of Olfaction and Gustation*, Marcel Dekker, 1995.

H. R. Moskowitz and C. B. Warren, *Odour Quality and Chemical Structure*, American Chemical Society, 1979.

*G. Ohloff and A. F. Thomas (eds), *Gustation and Olfaction*, Academic Press, 1971.

S. van Toller and G. H. Dodd (eds), *Perfumery, the Psychology and Biology of Fragrance*, Chapman and Hall, 1988.

PERFUMERY

D. P. Anonis, *Flower Oils and Floral Compounds in Perfumery*, Allured, 1993.

*R. R. Calkin and J. S. Jellinek, *Perfumery, Practice and Principles*, John Wiley, 1994.

L. Small, 'Perfumery: is it Art?', *Manufacturing Chem.*, Sept 1988.

C. S. Sell, 'Or Science?', *Manufacturing Chem.*, Oct 1988.

TERPENES

S. Dev, A. P. S. Narula and J. S. Jadav, *Handbook of Terpenoids*, 2 volumes, CRC, 1982.

T. K. Devon and A. I. Scott, *Handbook of Naturally Occurring Compounds, Vol. 2, The Terpenes*, Academic Press, 1972.

E. Klein and W. Rojahn, *The Configuration of the Monoterpenoids*, Dragoco, Wall chart and accompanying booklet.

E. Klein and W. Rojahn, *The Configuration of the Sesquiterpenoids*, Dragoco, Wall chart and accompanying booklet.

A. A. Newman, *Chemistry of Terpenes and Terpenoids*, Academic Press, 1972.

A. R. Pinder, *The Chemistry of the Terpenes*, Chapman and Hall, 1960.

Royal Society of Chemistry, *Terpenoids and Steroids*, Royal Society of Chemistry, annual review series.

J. L. Simonsen, *The Terpenes*, 3 volumes, Cambridge University Press, 1949.

*P. J. Tesseire, *Chemistry of Fragrant Substances*, VCR Publishers, 1993.

NATURAL PRODUCTS AND BIOGENESIS

J. D. Bu'Lock, *The Biosynthesis of Natural Products*, McGraw-Hill, 1965.

*J. Mann, R. S. Davidson, J. B. Hobbs, D. V. Banthorpe and J. B. Harborne, *Natural Products: Their Chemistry and Biological Significance*, Longman, 1994.

K. B. G. Torssell, *Natural Product Chemistry*, John Wiley, 1983.

BIOCHEMISTRY

*A. L. Lehninger, *Principles of Biochemistry*, Worth, 1993.

J. D. Rawn, *Biochemistry*, Harper and Row, 1983.

CHEMICAL INDUSTRY

A. F. M. Barton, *Handbook of Solubility Parameters and other Cohesion Parameters*, CRC Press, 1985.

Handbook of Chemistry & Physics, 69th edition, CRC Press, 1988.

**Kirk-Othmer Encyclopaedia of Chemical Technology*, 24 volumes + supplements, Wiley-Interscience, 1984.

H. H. Szmant, *Organic Building Blocks of the Chemical Industry*, John Wiley and Sons, 1989.

Subject Index